Origins
The Study Guide

The Origin of Matter, Space, Time, and Life.

Section 2 of 3

by Dr. Troy E. Lawrence
email address: Lawrence@creationministry.org
Send donations to: www.creationministry.org

Edited by George Macias

Copyright © 2016 by Troy Lawrence

Published by Troy Lawrence Publishing
ISBN: 978-1943185023

Printed in the United States of America.

All rights reserved. No part of this publication may be reproduced, stored in a retrieval system, or transmitted in any form or by any means—for example, electronic, photocopy, recording—without the prior written permission of the publisher. The only exception is brief quotations in printed reviews.

All Scripture quotations, unless otherwise indicated, are taken from the Holy Bible, New American Standard Bible®, copyright © 1960, 1962, 1963, 1968, 1971–75, 1977, 1995 by The Lockman Foundation. Used by permission.

Table of Contents

Preface   4
Introduction   4

Chapter   **Page**

Section II
How Old is the Earth?

12	Tectonic Plates and River Deltas	7
13	Radioactive Isotopic Dating	14
14	Carbon-14 Dating	22
15	Viscosity of Rocks	24
16	Moon Dust	25
17	The Magnetic Field	27
18	Polystrata Petrification and Fossilization	29
19	Distance to the Moon	37
20	Tyrannosaurus Rex Soft Tissue Found	39
21	Layers of the Earth	40
22	Transitional Fossils	54
23	Light	65
24	Humans Lived 900+ Years and Adaptation from Origins	70
25	What Happened to Dinosaurs?	83
26	The Big Bang Versus the Bible	93
27	The Big Bang Versus Physics	100
28	Evolution Versus Mathematics	109
29	Evolution Versus Physics	121
30	Evolution Versus Science	126
31	Evolution Versus the Bible	146

Preface

This study guide will help explain the origins of dinosaurs, humans, the earth, the sun, the moon, stars, galaxies, ice ages, polar ice caps, deserts, layers of sedimentation, petrification, fossils, vast oil reserves, and the oceans. In addition, this study guide explores the difference between evolution (macroevolution) and adaptation (microevolution) and the difference between evolution and creation. You will learn how dinosaurs became so large and why they are not visible today and what caused the dinosaurs to be extinct. We will cover such topics as whether it is probable for humans to live 900+ years of age and whether humans evolved from primates (monkeys)? Did dinosaurs and humans live together on the earth? How old is the earth? What is the origin of all living creatures? These questions and many more will be answered in these study guides.

The more knowledge and truth we possess, the less amount of faith is required to accept the testimony within the Bible. Accepting the records in the Genesis creation account actually becomes easier as knowledge is increased. This is in accord with Romans 10:17, "faith comes by hearing, and hearing by the Word of God." Thus, as you increase in your knowledge of the Scriptures, then your faith grows. The results for believers reading *Origins* is that their knowledge increases, which increases their faith, which leads to them reading the Bible more frequently and witnessing more. The results for nonbelievers reading *Origins* is that they see there is a superior interpretation of the observable evidence that is in harmony with science. Some reported results have included readers rejecting a major tenant of evolution, such as the notions that mutations enhance DNA to eventually form new functions and new kinds of creatures and that there is some Intelligent Designer.

This study guide should not be read by itself; it should be accompanied by the book *Origins* because the book contains more detail, footnotes, and additional evidence to solidify the points.

Introduction:
The Premise of *Origins*

One of the major components of this study guide is an explanation of how changes to Earth's gravity, oxygen concentration, rotational velocity, temperature, and the loss of a canopy of salt water that surrounded our atmosphere have adversely affected all living creatures on the earth—resulting in a severe reduction in the length of life, the size of life forms, and in the earth having four seasons, polar ice caps, deserts, and more.

The Motivation for *Origins*

There were several motivating factors that caused me to write *Origins*. The primary factor is that the reigning accepted hypothesis, evolution, is at odds with the Bible. Both cannot be correct, since they make polar opposite statements on the beginning of life. Secondarily, there are too many people that have no idea what to think or believe regarding the origins of life and all things.

Origins is such an important topic because how people view the origins of life shapes their entire outlook on all faiths, hope, security, the source of truth, and the purpose for life. There are basically two views—and it's not science versus God, it is evolutionary scientists versus science and God—regarding the origins of life:

(a) The Big Bang initiated all processes and laws, and all life forms evolved from a single-celled organism that spontaneously began from a primordial complex chemical mixture into the complexities of life we see today by explainable natural processes in nature.

(b) God created everything by natural and supernatural processes in six rotations of the earth.

All of public academia, from elementary school age to graduate level education, teaches evolution. And almost all media teach evolution. This teaching is so ubiquitous that even 90% of seminary schools and most churches teach an old Earth time line to coincide with the evolutionary model. Some churches say that evolution is truth, but God is behind it all. The overwhelming majority of people on earth say that the Bible requires too big of a leap of faith to believe that it's 100% truth; it is just a good moral book. Even the Pope has accepted the old Earth hypothesis and evolution. When broad is the way and many follow, that should give us pause (Matt. 7:13–14).

The serious implications of the origins of life are striking. Start with the wrong premise, and conclusions built on that premise will be in error. If evolution is correct, then the Bible is not from God, but from men and filled with errors. If the Bible's account of creation is correct, then evolution is error. There is no middle ground, since the two views are so opposite. There is no room for compromise. If one verse in the Bible is wrong, then it's not a book from an all-powerful God; it's a book from fallible humans. Frankly, it would need to be burned. But if the Genesis creation account is correct, then evolutionary teaching and doctrine need to be exposed and stopped.

Why does it have to be so cut and dry? Why can't there be tolerance of both views? Because the Bible takes a very hard line and proclaims that there is no middle ground. Jesus said, "He who is not with Me is against Me; and he who does not gather with Me scatters" (Matthew 12:30). Therefore, there is a lot at stake at the very beginning of the origins of life because the Bible declares God cannot lie (Titus 1:2), so if the Bible is in error, then God is a liar and the Bible is made up by man.

An honest scientist would say that evolution and the Big Bang are accepted hypotheses, but invariably when professors lecture, they say evolution is fact. The prevailing, leading, and accepted view is that evolution and the Big Bang are not just possible hypotheses that are further being explored as good theories: they are facts. The implication is that if the Bible's creation account has one error, it is terminal. But on an individual level, people are less likely to read, study, or memorize a book they believe contains errors. Thus, they are less knowledgeable on how to be obedient to God, and by default, more prone to erroneous activity, thought, and doctrine. And the subsequent result is that they are less likely to share truth with others because no one wants to approach an atheist and say, "believe in my all powerful God, though He wrote errors in Genesis creation."

Are there holes in the evolutionary hypothesis? Is evolution fact? If there are holes in the evolutionary model, there is too much ignorance amongst Christians to know them and too much ignorance about what the Bible teaches for them to defend the Bible. Most people don't know what to think or believe regarding the origins of life. They merely accept whatever they are told. This bothers me for two reasons.

Reason No. 1: The Bible declares that God wrote the Bible through mankind (II Timothy *3:16)*. Therefore, if there is one error in the original manuscripts, for example, in such things as the Genesis account of creation, then the whole Bible can't be infallible and be written by God. It should be viewed simply as a good book along with any other good book. But if the Genesis account of creation is the truth, then indeed the Bible is the Word of God, and evolution and the singularity of the Big Bang may represent the fulfillment of I Timothy 4:1 "But the Spirit explicitly says that in later times some will fall away from the faith, paying attention to deceitful spirits and doctrines of demons."

Reason No. 2: II Timothy 4:1 says,

> I solemnly charge you in the presence of God and of Christ Jesus, who is to judge the living and the dead, and by His appearing and His kingdom: <u>preach the word; be ready in season and out of season</u>; reprove, rebuke, exhort, with great patience and instruction.

Ignorance amongst Christians on the subject of origins is disobedience to the Word of God and is sinful. Christians are at war spiritually for the truth, and the Word of God is the sword (Ephesians 6:12–17). To be ignorant of what the Bible says about the beginning is an act of willful negligence. To claim to be a Christian and accept evolution as truth and to believe the Bible to be in error is an act of spiritual treason. There is no neutral ground; Christ said you are either for me or against me.

Of those that believe in God, there are too many that just don't have a clue about what happened in the beginning. And this certainly renders them catatonic when an opportunity comes for them to discuss the difference from the Genesis account of creation and evolution. Imagine how ineffective a Christian becomes at witnessing when they tell others to believe in Jesus and His Word, yet they themselves don't even believe Jesus' Word as truth. Why would anyone believe in a God of the Bible that doesn't speak truth regarding the origins of life? They wouldn't.

The battle for the beginning is extremely critical. And for this reason, Satan attempts to separate humans from the Word of God before tempting. Consider Adam and Eve; Satan said, "Indeed has God said," and then he tempted them (Genesis 3). And after Jesus was baptized, anointed by the Holy Spirit, and heralded by the Father with, "This is My beloved Son," then Satan attacked the Word of God by saying, "If You are the Son of God," and then tempted Him (Matt 3:17– \4:9). Therefore, we may conclude that Satan understands the best way to tempt someone to sin, is to first attack the Word, separate people from the Word, then they are easy prey to lead into sin. And there is no better way to do this then to start with *In the beginning*. We have seen the world reject God's creation account and accept man's hypothesis of the beginning and the exponential rise of sin as a result.

Let's journey together through the sciences, logic, and the Bible to determine what is fact or fiction, what is truth or error, and what is from God or from man.

Group Discussion:

1. Have you heard evolutionists proclaim their view as fact and how did that impact you?

2. As a result of the dominant view of evolution, has this limited the amount of time you have read the Bible, or shared the gospel?

3. Have you ever felt defenseless to defend creation and how did that affect you?

4. If this book proves that the Genesis creation account is true, and the earth and universe are young, how will that effect your reading and sharing of the Bible?

Section II

How Old is the Earth?
Chapter 12
Tectonic Plates and River Deltas

Scientists theorize that at one point in time, the earth's continents were connected together into one place and the seas/oceans were in one place surrounding the land. The seas that were gathered in one place were called Panthalassa, and the land that was gathered in one place was called Pangaea.

When one looks from an aerial view of the earth, it doesn't take long to see that indeed the continents fit well together. But that's only a partial view, for beneath the water line, beneath where the ocean meets the shoreline, is the continental shelf. The continental shelf extends around the continents and was the shoreline during the ice age. Also, the polar glacial ice was so vast in the months following the Flood that it reached half way to the equator. With the polar ice caps being so large during the glacial age, they would have lowered the flood levels and formed a lower shoreline that we call the continental shelves. When one takes a look at the continents and the continental shelves, one can see that the continents fit even better together with the continental shelves than with the actual coastlines.

The Bible implies that the continents are connected in Gen. 1. On the first day of creation, God creates matter, space, motion, time, and light. The burgeoning planets were already rotating on an axis, hence the phrase of evening and morning and the first day. On the second day of creation, God stretches out the universe (stars, planets, and moons are still forming at this point) violently, and this forms our atmosphere. Now, when God created the atmosphere, He separated the waters from the waters to create an expanse, or an opening in the midst of the waters, which puts water above the atmosphere and water below the atmosphere (Gen. 1:6). At this point in creation, the earth was surrounded by a spherical canopy of water and atmosphere, thus suggesting the earth was spherical, but there was no defined form to the land, as the waters were mixed with the dirt.

Gen. 1:9: *"Then God said, 'Let the waters below the heavens be gathered into one place, and let the dry land appear'; and it was so. God called the dry land earth, and the gathering of the waters He called seas."* In the Strong's Concordance, the words *one place* means one "place/one spot" or "like kinds of waters were gathered" (saltwater and freshwater lakes and deep freshwater reservoirs). This indicates that the seas were together. And while moving the seas around into one place, at the same time God said, "And let the dry land appear." Logically, if the seas are gathered into one place, then the land also has to be gathered into the other one place, that is, Pangaea. How God separated the water from the land, was simply by what geologists call post-glacial rebound (or continental rebound). With the weight of the universe removed from pressing upon the earth's dirt, the land expanded with hills, elevations, topography, and so forth. Water simply flowed to lower elevations.

When did Pangaea break apart? Well, we need to lay a foundation for a few things before we can dive deeply into when Pangaea broke apart. First, we need to know that the continents are sitting on tectonic plates. And these tectonic plates move, causing the continents to move. There are divergent zones where the crust of earth is splitting apart pushing the large continents away from a central origin. The Mid-Atlantic Ridge pushes the North American and South American plates toward the Pacific Ocean, causing multiple subduction zones, where one tectonic plate is sliding underneath another tectonic plate. This causes the "ring of fire," a ring around the Pacific plate where it meets the North American and Asian plates; it is the source of many earthquakes and volcanic activity.

Evolutionary geologists estimate that Pangaea broke apart some 120–220 million years ago. How do they come to this conclusion? They look at the distance traveled (for example, between South

American plate to African plate is 2848 km/1,770 miles), divided by the current rate of tectonic plate speed (2.5 cm/yr). This is roughly 115 million years ago. But wait a minute. We don't need to know the distance from South America to Africa; we need to know the distances from South America to the Mid-Atlantic Ridge (sight of origin) and from Africa to the Mid-Atlantic Ridge. And that distance is half the amount, or 885 miles (1424km), which equates to ~57 million years.

With this calculation, old earth evolutionists and young earth creationists seem to be in a pickle. How do young earth creationists, who believe the earth to be created in six days around 6,000 to 10,000 years ago, wiggle out of this jam—by falling back on the scientific method (observation and testing) or just saying the Bible tells me so?

There are moments when tectonic plates move quickly, allowing for a hypothesis that it is possible for tectonic plates to move quicker than 2.5 centimeters per year, which opens the door for a younger Pangaea breakup. In 2011, Japan had an earthquake that moved a GPS station 8 feet closer to Alaska in only two minutes. This equates to 0.045 mph; that's a velocity of 10 million centimeters per year as compared to the current projected velocity of 2.5 centimeters per year.

With a fast-moving tectonic plate hypothesis during the Genesis flood, we would expect to find evidence that there was a great deal of stored up potential energy in the tectonic plates, and we would expect to find some evidence that the friction that slows tectonic plate movement was substantially reduced during this fast tectonic plate event. In addition, we would expect to find evidence that the tectonic plates are capable of storing potential energy, and can suddenly release the energy, causing fast movement. And that is exactly what we observe and can measure. There are remnant left-behind deep reservoirs of water that didn't get squished out by the weight and movement of the tectonic plates during the Flood. In several places around the world, scientists have found freshwater reservoirs, such as underneath Israel, which opens the door to a hydroplaning hypothesis (by Dr. Walter Brown, author of "In the Beginning"). In addition, geologists affirm that tectonic plates are capable of storing potential energy and suddenly releasing the stored up energy as fast kinetic energy, which allows the hypothesis that potential energy was stored up from creation to the Flood. If there were no tectonic plate movements from creation till the Flood, there would be ±1,650 years of potential energy stored up, which is a sufficient amount of potential energy to account for the fast-moving kinetic energy hypothesized with a quick plate movement during the Flood. And since the term *fast tectonic plate movement* is relative, we are only talking about 2.5–5 kmph (1–2 mph) during the yearlong Flood saga. To put that in perspective, humans walk at 2.5 mph. This is certainly plausible. One earthquake in 2011 caused the tectonic plate that Japan is on to move 164 feet in minutes, and the 2004 Sumatra earthquake shifted a tectonic plate ±80 feet. But what is different about the plate movements during the Flood is that they were hydroplaning on the fountains of the deep to reduce the drag/friction coefficient. Whether the hydroplaning of the tectonic plates were on literal water or on highly viscous liquid lava is immaterial because scientists have discovered that volcanic eruptions emit a large quantity of water into the air. When volcanoes erupt, they do appear as fountains from the deep, within the earth. And since volcanoes often contain a lot of water, it is possible that the fountains of the deep mentioned in the Flood saga were highly viscous and volatile volcanoes. I suspect the answer is both, liquid water bursting from the fault lines as Dr. Kent Hovind and Dr. Walter Brown contend and highly viscous volcanoes containing water. Even today, scientists still find an abundance of water mixed within the matter of the mantle.

We observe today that the tectonic plates are moving slow enough to allow river deltas to form at the mouths of rivers. Then there should be at least one trail of sediment on the ocean floor from where the rivers once laid sediment to the current mouths of rivers. If the tectonic plates have always traveled at this rate of speed, and the rate is slow enough now for deltas of sediment to form, then there should be prior evidence of river delta formations along the ocean floor representing millions of years. One may argue that this is impossible because the mouths of the rivers are traveling with the tectonic plate and therefore, there cannot be a trail. But this is only a half truth because there are locations around the globe where fault lines intersect rivers and their deltas, and thus there should be indicators of prior deltas that establish an existence greater than 4,500 years. For example, on the Northern California coastline, there are several rivers that empty into the Pacific ocean, and their sediment deposits cross over the San Andreas fault. But guess what? There are no detectable trails of sedimentary deposit from any river, and more importantly, there is no river delta of sediment representing greater than 5,000 years of existence. The implication is that the tectonic plates have moved too quickly for rivers to deposit a sediment trail on the ocean floor as the continents have drifted and that the date of Pangaea breaking apart is relatively recent.

Psalm 104:5–9 provides a summary of Earth's history; after the foundation of the world (creation), God caused the deep (the Flood) to cover the entire world, and then He raised the mountains and lowered the valleys. And this is what it says, "He established the earth upon its foundations, So that it will not totter forever and ever." That is the third day of creation, at the foundation of the world. Then it says, "You covered it with the deep as with a garment; The waters were standing above the mountains." This is the catastrophic global flood: "At your rebuke they fled, at the sound of Your thunder they hurried away." God rebuked the floodwater, and there is no mention or association of God rebuking any portion of the water during the creation moment of Gen. 1–2 because all that God made was very good (Gen. 1:31), so this is after the rain stopped on the 40th day and after the water had remained and prevailed above the earth for 150 days. Psalm 104:8: "The mountains rose; the valleys sank down to the place which You established for them." This verse aids in the clarity that the tectonic plates were still moving from the breakup of Pangaea at the start of the Flood. They had drifted quickly across the surface of the earth and had now reached contact with other tectonic plates. The fast-moving tectonic plates were traveling at ~1–2 mph as they impacted with other plates, causing collisions that buckled the crust of the earth and sent mountains higher and valleys lower. The tectonic plates buckled either up or down to alleviate the force, just like a car in a crusher. After the rain and the waters prevailed for 150 days, God caused the mountains to rise above the surface of the waters and caused the valleys to sink down, setting the boundaries to prevent another global flood. Psalm 104:9: "You set a boundary that they may not pass over, So that they will not return to cover the earth." The brevity in terms of how quickly the tectonic plates moved to achieve this explains why the Himalayas are so tall yet have seashells on them. It explains where the waters went, and for this discussion, it explains why there are no trails of sediment deposits on the ocean floors.

As a side note, evolutionists are quick to say, "There is not enough water on earth to cover the high Himalayas, so the Bible legend of the Flood was regional or didn't happen." This is where Hugh Ross is wrong, and he has led many people astray with his error. Well, the mountains we see today are much taller than when the Flood occurred, and the valleys we see today are much deeper than when the Flood occurred, as explained in Psalm 104:5–9. In fact, if the spherical Earth were smoothed out with no mountains, no valleys, and no polar ice, then the entire planet would be covered ~1.5 miles deep with water. And Gen. 7:19–20 says every mountain was covered 15 cubits (22.5 feet) by water, not a mile, so the earth wasn't void of mountains and valleys, but they weren't as tall as we see today. A global flood explains why there are petrified clams with closed shells on top of the Andes mountains. Clams lose motor control with death and cannot keep their shells closed; therefore, they typically open up their shells when they die. Thus, the closed shells suggest that they were quickly covered with soil, moisture, and pressure as the Genesis flood provides, and this prevented the shells from opening.

Additionally, before the Flood, there was a canopy of water that hovered around the atmosphere. This canopy significantly raised the atmospheric pressure and prevented air moisture to form into clouds for raindrops. The Bible helps with this information by explaining that a mist rose from the ground to water the whole earth (Gen. 2:6), and the first cloud and the first rainbow came after the Flood, when atmospheric pressure was reduced to current levels. What does this have to do with having enough water to flood the earth? Well, before the Flood, the atmospheric pressure was too high to have moisture in the form of clouds. And today, on average, the atmosphere contains 37.5 million billion gallons of water as vapor. That is enough water to cover the entire surface of the earth with one inch of rain.

At some point, when one sees that evolutionary hypotheses are not fact and that no record in the Bible is wrong, then one either has to have faith in evolution or accept the 100% accuracy of the Scriptures. There will be a time in the future when every island and every mountain will be removed (Rev 16:20), and an asteroid will hit the earth so hard that it will split the earth asunder, and it will wobble like a drunkard (Isaiah 24:18–20) and lead to the disruption of the seas (Rev 21:1). Therefore, if one struggles with the verses of the Bible that recorded the past and then changes the Scriptures to fit man's hypotheses, then one might as well throw the whole Bible out. Greater things are coming soon, so one must choose either to fully accept the evolutionary hypothesis or the truth of the Bible as it is. But one can't accept some of the Bible as truth and evolution as fact.

Evolutionists may argue that since oceans are not static, maybe the currents moved the sediment away. There is no disputing that oceans are dynamic and have currents. But ocean currents near the ocean floor, that have allegedly swept away all the sediment trails are 1/100 the speed of the surface currents. Therefore, the argument that the sedimentary delta trails along the Pacific Ocean floor (where the San Andreas fault intersects many river deltas) disappeared because of turbulent currents is invalid. The current at the bottom of the ocean is 0.044 mph (0.02 m/s). A human walks 56 times faster than ocean currents near the ocean floor.

A river delta is a fan-like shape of soil deposits at the mouth/opening of a river. And each river has a delta with deposits of sedimentation. But since Pangaea allegedly broke apart 120–220 million years ago, there should be a river with about 120+ million years worth of sediment deposits to confirm this belief. However, there is no river on the face of earth that has enough sediment deposit to establish that Pangaea broke apart 120 million years ago. In fact, there is no river on earth that has enough sediment deposit at the mouth of the river to establish that Pangaea broke apart 10,000 years ago. The

amount of sediment deposited only supports that Pangaea broke apart roughly 4,500 years ago.

The image shows the continents and the ocean floor. Once a river's sediment deposits reach a high enough level, an island would form and a bridge of sediment would connect the mouth of the river with the newly formed island of sediment. The river delta would have to move to a new location to form another free-flowing spot to deposit its sediment. Now, there are no multiple river sediment locations to equal enough years of existence to establish greater than 10,000 years. In fact, each river's sediment deposit only has enough sediment to establish about ±4,500 years of existence. Unless evolutionists want to start arguing that there was no rain to cause erosion for 220 million years, then they are stuck with an age that doesn't match the observable evidence. *Image credit: www.colonial.net/Plate tectonic theory.*

Combining the information together, there is one conclusion: Pangaea broke apart around ~4,500 years ago during the Gen. 7 global flood. During that chaotic event, the tectonic plates moved too quickly for a trail of river sediment to form on the ocean floor as the continents drifted apart from Pangaea. And ever since the Flood of roughly 4,500 years ago, the continental tectonic plates have been moving slow enough (1.5 in/yr, 2.5 cm/yr.) to allow river deltas to form.

How does the Bible handle the breaking apart of Pangaea? As we discussed earlier, the concept of Pangaea is in harmony with the Bible. The Bible explains that God created everything in six days about 6,000 years ago. This can be determined by taking a literal (plain) reading of the Genesis account and utilizing its genealogies. Then a major catastrophic event occurs around 4,350 years ago, just 1,650 years after creation. But many people miss the importance of one excerpt from Gen. 7:11: "The fountains of the great deep burst open." The Bible gives us an indication that something powerful forced water to burst violently out of deep caverns from under the crust of the earth, which indicates that the tectonic plates may have been hydroplaning on a bed of water. This would make the tectonic plate movement easier and quicker and would have forced water and soil to burst out of the deep caverns. Science gives us some help in identifying the possible kick-starter of this global catastrophe—asteroids.

Seismologist Tom Jordan of the University of Southern California explains that adding fluid to the crust of the earth "decreases the strength of the rock, and therefore they break," causing earthquakes and faults. Adding water to the crust of the earth is exactly what the Flood accomplished to weaken Pangaea's fault lines.

Imagine large asteroids passing through the canopy of salt water that once hovered above the atmosphere. The asteroids would have broken up the external frozen arch support of the canopy, causing the water to start to come down upon the earth as rainfall. The asteroids impacted Pangaea, fracturing the crust, releasing 1,650 years of stored-up potential tectonic plate energy, and setting in motion the events that split Pangaea into multiple large fragments that we call continents. The fragments moved away from what was Pangaea, causing openings to the great fountains of the deep, with the weight of the fractured crust squishing down upon the deep caverns, causing water to burst violently out of the great fountains of the deep. The asteroid impacts sent tons of iridium ash and soil into the air, which settled upon the seven new continents.

One thing to remember is that fossils, petrification, coalification, and petrolification do not need millions of years to form as we have been told. They need botanical or biological specimens to be covered quickly with soil, moisture, heat, and pressure, within a brief amount of time. All the fossils, petrified wood, coal, and petroleum were formed as a result of the Flood because they all represent that something died and was covered quickly. The Bible teaches that there was no death prior to Adam and Eve sinning. And since there is no other catastrophic event prior to or after the Flood, this means that there were no fossils, petrified wood, coal mines, or oil reservoirs prior to the Flood. All the fossils indicate that they were quickly covered because all things decay to dust in a brief moment in time. For leaves, they decay to dust within months. For creatures, they decay to dust within years. Thus, every fossil is testimony against a slow deposit hypothesis from evolutionists and is positive testimony for the Bible's catastrophic flood record.

The Bible gives us some clues that Pangaea broke apart ~4,500 years ago, but we get additional information when the broken Pangaea and the subsequent seven continents were no longer connected/accessible by traversing along the continental shelf or ice pack. This information is in the genealogies of the Bible. Noah's son Shem begot Arpachshad (two years after the Flood, Gen. 11:10–18), Arpachshad lived 35 years and begot Shelah. Shelah lived 30 years and begot Eber. Eber lived 34 years and begot Peleg 101 years after the Flood (each father had other sons and daughters). The only way to move to different continents was by traveling along the continental shelves, by ice, or by boat. The name Peleg means divided, to split, earthquake, son of Shem (Great Grandfather of Peleg), and "for **in his days** the earth was divided" (Gen. 10:25). This puts Peleg's life during the time of King Nimrod's fatal decision to build the Tower of Babel, which led to the scattering of the people around the globe. Thus, the moment of when the Tower of Babel occurred was during the days of Peleg, not at his birth 101 years after the Flood.

What was the Tower of Babel? Noah's great grandson, King Nimrod, wanted to make a name for himself, so all the people with one language and one accord built a tower to reach the heavens to become like God. Well, God came down and confused their languages and scattered them abroad over the face of the whole earth.

When looking at the Hebrew definition for *Peleg* (#6389) in the Strong's Concordance, there is an alternate definition right above *Peleg*. #6388 *peleg*: a rill (i.e., a small channel of water, as in irrigation), river, or stream.

The alternate definition could shed some light on the second century after the Flood, as the ice would be melting gradually each day, and the ocean levels would be rising from small channels of water, rivers, and streams from the melting ice. My contention is that Peleg's name provides more information regarding the time of the Tower of Babel, when God confused their languages and how God scattered them abroad over the face of the earth. People literally walked along the continental shelves and the ice to the different continents as God motivated them to do so. In other words, God used natural means to carry out His supernatural miracle of scattering the people across the globe. When the waters rose above the continental shelves, then the people were no longer connected because enough of the polar ice caps had melted to raise the ocean levels to cut off the routes of passage between the continents—thus, finally making England and Australia as islands.

During the time of the Tower of Babel, God changed people's languages, and then they literally walked to their new locations, establishing different cultures with different languages. And people adapted to climate change, new diets, and diseases. Skin colors changed in some parts of the world due to adaptation.

To be fair, I have read that some pastors interpret the dividing of the people and the earth at Peleg's birth differently. This alternate view of the day when Peleg was born is accompanied with the concept that a great earthquake ripped Pangaea apart. It's not a well-supported interpretation because then that weakens the argument in terms of the destructive power of the global flood, and then the mechanism that forced water out of the deep caverns is gone. And Psalm 104 declares that when the Flood covered the earth, then God rebuked the waters and caused the mountains to rise and the valleys to sink. That's why I like the melted ice theory, in which enough ice had to form to start the glacial age, which caused ocean levels to decline, and then as ice melted, this caused ocean levels to rise above the continental shelf during the days of Peleg, thus dividing the lands. This event did not occur at the time Peleg was born because 101 years doesn't seem enough time to melt significant amounts of ice to raise the ocean levels above the continent shelf, and the Bible clearly says, "For in his days the earth was divided," and it does not say, "at his birth the earth was divided."

Names of individuals back in the patriarchal times were prophetically given based on events that would occur or the nature of that person's life. Today, we may not fully fathom the influence of the Holy Spirit guiding parents to name their children, as when God, Emmanuel (Jesus), physically walked with mankind. To understand how important names were and really still are today, consider the meaning of the names of the people in the Bible in order:

Name (Strong's Concordance). Definition.
Adam (# 120)   Mankind.
Seth (# 8352)   Appointed (to), substituted, corresponding to.

Enosh (# 583)	Mortal.
Kenan (# 7015–7018)	Sorrow, lamentations.
Mahalalel (# 4111)	The Blessed God.
Jared (# 3382)	Shall come down, a descent, descendant, an heir.
Enoch (# 2585, 2596)	Discipline, dedicate, train up, teaching.
Methuselah (# 4967)	His death shall bring [Methuselah died the year the Flood started].
Lamech (# 3925, 3928)	Disciple, instructed, learned the Despairing.
Noah (# 5146)	Rest/comfort.
Shem (# 8034-5)	Prosperity, a memorial of individually by honor, authority, character.
Arpakshad (# 775)	of for.
Shelah (# 7974-5, 7999-8004)	Fountain of peace, peace, Jerusalem, make amends, make restitution.
Eber (# 5676-7)	A region across, on the opposite side (of water), passage, other side.
Peleg (# 6388)	a rill (i.e., a small channel of water, as in irrigation), river, or stream. "for in his days the earth was divided."

Putting the meanings of the names together forms a prophetic message for all of creation: mankind, appointed to mortality, and sorrow. The Blessed God shall send an heir teaching. His death shall bring the despairing disciples rest by His authority of Peace across the regions that are divided by water.

The point of the names for this discussion, besides the Gospel message, is that the names of the people seem to be for future events, not past events. Thus Peleg, though his name means divided, doesn't equate to "at his birth the earth was divided," but to "in his days the earth was divided." Therefore, the time when enough ice melted to raise the ocean levels above the continental shelf occurred during Peleg's days, sometime after the Tower of Babel. This means that the people of the Tower of Babel were able to travel to distant continents and gradually adapt to the new environment and become all the different cultures and races we see today.

Chapter review: The evidence suggests that the tectonic plates moved quickly during the Flood, and that explains why there are no sedimentary delta trails. In addition, there is no river on Earth that has enough sediment at its delta to establish an age greater than 4,500 years. This too indicates that Pangaea broke apart recently. The glacial age was still receding four generations after the Flood, during the days of Peleg's life (not at his birth), and people were scattered around the globe because of the sin of that occurred at the time of the Tower of Babel. Shortly after the people had traveled to distant continents, enough ice had melted to raise the ocean levels above the continental shelves. Their travels were cut off when the ocean levels rose, until technology would connect mankind again.

Group Discussion:

1. With the oldest river delta being ~4,500 years old, how does that influence your faith in the Bible?

2. Knowing that tectonic plates can move fast at times, how does a release of 1,650 years worth of stored potential energy, from creation to the Flood, influence your understanding of when the continents broke apart and the catastrophic Genesis Flood?

Chapter 13
Radioactive Isotopic Dating

We are told the earth is 4.5 billion years old. This is taught from elementary school all the way to graduate level education. All secular television shows teach an old-earth paradigm and mock young earth creationists. All government-supported institutions teach an old-earth belief system as well and have fired "rogue" young earth creationist professors for speaking up. And 90% of all seminary schools teach an old earth as well. Is this accurate? Are all of them correct? Let's find out.

Let's determine the best means that the evolutionary scientists use to determine the age of the earth. The best and most accurate method is radioactive isotopic dating, which is a simple algorithmic (math) formula. Radioactive elements are unstable in their natural state, and they want to become stable. The way they become stable is that they lose electrons. The first stable element on the periodic chart is hydrogen (H), and the last is lead (Pb). Lead isotopes are the natural end products of several radioactive isotopes (such as polonium). The rate at which an unstable radioactive element loses an electron is measurable. Therefore, we can look at the number of stable lead ions compared to the number of unstable polonium (Po) ions and determine how long the unstable element has been in existence and losing ions to become lead.

The simple math is to take the number of lead ions (the last stable element) divided by the number of polonium ions (the first radioactive element) multiplied by the **CRD ("Constant Rate of Decay,"** the time it takes for a radioactive element to lose an alpha particle in an attempt to become more stable). The more alpha particles of electrons that have left the radioactive element, the older the object is that is being tested. *Image credit: Wikipedia/Periodic Table of the Elements.*

So where's the problem with this formula? There are two problems. One is that there are examples in nature and in granite rock on every continent that show some polonium halos in a primordial (first developed/original condition) state with no parental halos. Meaning that there is no evidence that the polonium descended from heavier elements. Second is that the special multiplier, the CRD is not constant. Of course, the formula is only accurate if the CRD has always been constant. No scientist lived 5,000 years ago to test or examine the rate of decay back then. Scientists can only determine that since they have been able to quantitatively measure the rate of decay, it seems to be a constant in our limited observation. Since there were no ion mass spectrometers invented till modern times, then no scientist can definitively say that the rate of decay has always been constant. Therefore, it is a leap of faith to say that the primary way of determining the age of the earth is by multiplying by a rate of decay that we perceive to be constant today, when we have no assurance that it has always been constant. If the rate of decay has not been constant, then the entire algorithm formula is wrong, and the age of the earth is wildly inaccurate and potentially younger by billions of years.

This rate of decay is also the foundation of evolution. Since evolution needs billions of years for life to evolve, and radioactive dating tells scientists that the earth is billions of years old, evolutionists need the rate of decay to be constant. If the rate of decay has moments of acceleration, then the foundation of evolution is gone because the foundation of billions of years is gone. Evolutionists would be in a pickle if life on earth was only 6,000 to 10,000 years old.

Evolutionists may want to validate their point by saying that the rate of decay may have had moments of deceleration, thus giving evidence that the earth and life on earth is much older than anticipated. There are examples of deceleration of decay, such as in a deep freeze. Freezing things, as we know, slows the decay process. But deep freezes also slow growth, and by definition, would slow evolutionary changes. If an evolutionist wants to employ the deep freeze theory as evidence to counter the argument of accelerated rates of decay, then they are using an argument that weakens their own belief system. In a deep freeze, their Darwinian evolutionary systems grinds to a halt. Therefore, any deep freeze on the planet to explain a theory of an older earth hinders the evolutionary process and thereby gains no ground for the argument. Also, when an evolutionist says that there were moments of decelerated rates of decay, then they, by default, validate my argument that the CRD is not constant, and there were/are moments of accelerated rates of decay.

Are there any examples of acceleration in the rate of decay that support a creationist's hypothesis? Yes. Imagine twins born and separated at birth for this aging experiment. One lives in the Sahara Desert (more of the sun's radiation), with its dry air and hot temperatures depleting moisture from the skin with each passing day. Making matters worse, this twin enjoys laying out in the sun to get a tan. Now let's add some vices to this twin who lives in the desert. Let's add smoking cigarettes and drinking alcohol. And also for good measure, let's say that this twin works long hard hours at its vocation, which causes stress and loss of sleep.

The other twin lives in a mild climate with high humidity and protects his or her skin from the sun's radiation. And this twin's lifestyle consists of healthy eating, exercise, sleep, and an easy job. On the twins' 60th year birthday party, they are reunited. Though the twins are the same age, with the same genetic makeup, the rate of decay was not constant for both of them. The twin that lived in the desert would look older than the twin who lived in the mild climate. Why? The aging process was accelerated.

But we are talking about the earth, not skin. Are there examples of the rate of decay not being constant for the earth or objects on the earth? Yes. Dr. Robert Gentry has done great work on this. He has taken a piece of wood and subjected the wood to trauma. He sealed it in a vacuum (tube with no air) with trace elements of clay and water. He baked it for eight months at 150°C (mimicking a buried earth scenario next to a volcano to represent the "trauma"), and that piece of wood changes to 100% coal. What is interesting about this experiment is that scientists have purported that coal forms after 20 million years from decaying wood. However, laboratory-created coal is indistinguishable (Nature, page 316, March 28, 1985) from the coal formed naturally by all the techniques so far applied to it. Scientist at Argonne National Laboratory have produced coal from natural materials in less than one year that is indistinguishable from naturally formed coal by all the techniques so far applied to it (Chemical & Engineering News. November 21, 1983. Page 42). The eight-month-old coal was indistinguishable from 20-million-year-old coal found in mines (Dr. Robert Gentry, Halos.com/videos, "Finger Prints of Creation," and "The Young Age of the Earth").

The implication is that trauma (from heat and pressure) accelerated the aging process of wood to coal. A process that allegedly takes 20 million years to occur, took only eight months. This is another example that the "constant" part of CRD is not constant. This is evidence that the age the earth, as scientists tell us, could be off by billions of years.

Are there examples in nature of trauma accelerating the aging process? Yes. Mount St. Helens, Washington, had a violent volcanic eruption in 1980. The massive trauma resulted in the petrification of trees in only three decades. Petrification is the process of a tree turning into stone. Scientists tell us that this process takes 500,000 years. For petrification to occur, there are four basic requirements: (1) trees that are quickly covered with soil (this must be accomplished before the tree decays to dust—~within a couple of years), (2) moisture, (3) pressure, and (4) heat and a source of silica. Why did some trees petrify in 30 years, when scientist tell us it is a process that takes 500,000 years to occur? It was because the violent trauma from the volcanic eruption accelerated the aging process. Scientists base the rate of decay on what we see today and extrapolate that the rate of decay today was also the same rate of decay forever in the past. This is a leap of faith. Some argue that the Mount St. Helens eruption in

1980 only uncovered petrified wood from prior eruptions that converted wood to stone. Well, then how far back do they want to speculate as to which prior eruption caused the wood to turn to stone? And therein lies the crux of the matter, for if they speculate that the 1980 eruption didn't produce petrified wood but only uncovered the eruption of 1847, 1800, 1500s, or 1400s, this still validates my point that trauma accelerates the aging process. For any of those dates doesn't come close to the 500,000 year mark that evolutionary geologists purport that it takes for petrification to occur.

Another example of the rate of decay being accelerated is in the conversion of carbon-based organisms into petroleum. In today's society, not much goes to waste. Even parts of a chicken that are not socially accepted as edible are not wasted. Chicken byproducts, such as adipose, ligaments, veins, and others, are converted to usable products, such as oil. The chicken byproducts are heated, and through a process of distillation, in 30 minutes, petroleum oil is created.

We are told that over a very slow decay process, organic material, such as algae and animals, decayed for millions of years to form petroleum some 50 to 300 million years ago. Yet, no scientist was ever there to confirm this huge length of time. And evolutionists won't tell you that chemists have figured out how to convert organic material to oil in 30 minutes.

At this point, you may be starting to get a little disappointed that evolutionists tell us petrification takes 500,000 years, coalification takes 20 million years, and petrolification takes millions of years, but they won't tell us that nature produces petrified trees in 30 years and mankind can produce coal from wood in eight months and petroleum oil from organic material in 30 minutes. Our property tax money goes to fund this unproven hypothesis of earth's age of billions of years, and our unsuspecting children hear this one-sided hypothesis, and now the churches are teaching this nonsense. Whether we are discussing radioactive decay rates, cellular decay rates, molecular decay rates, and so forth, trauma accelerates the decay process. Trauma, whether mechanical, chemical, or radioactive energy, accelerates the rate of entropy. Thus, when radioactive isotopic dating determines something to be hundreds of millions of years old, that is based on a leap of faith that the decay rate has always been constant, as nature and mankind have proven that decay rates are not constant. A German lab, Physikalisch-Technishe-Bundesanstalt, confirmed that uranium-226 has seasonal and monthly variations based on solar activity, thus, proving the CRD is mathematical trickery. And considering that lightning can ionize elements, then as Pangaea was breaking apart, and tectonic plates were sliding across the surface of the globe during the Flood saga, then there were massive amounts of electrical energy that accelerated the decay rates of radioactive elements. The violence by which God created all the matter on the first day and then expanded the universe on the second day with a Big Bang—those two days were so extremely violent that they both trump the energy released during the Genesis catastrophic flood. Since trauma accelerates the rate of decay, it is no wonder that the age of things appears older than their chronological age.

The reality is that when there is trauma, the rate of decay is accelerated. Therefore, the alleged best and most accurate means of testing the age of the earth and the things on the earth is wrong. Radioactive dating tell us that things are far older than they really are. Comparing the notion that Mount St. Helens' volcanic eruption produced petrified trees in 30 years to the notion of radioactive dating telling us it takes 500,000 years to produce petrified trees, suggests that radioactive dating could be off by 99.99%. What are some traumas that accelerate the aging process of the planet and life on the earth? They include asteroid impacts, fires, earthquakes, landslides, volcanoes, pyroclastic flows, floods, comet impacts, lava flows, tsunamis, tectonic plate activities, and extreme heat, all of which occurred during the Flood. In addition, there are hurricanes, tornadoes, solar flares, coronal mass ejections (from the sun), ultraviolet light, infrared light, visible light, gamma rays, x-rays, and droughts, all of which occur on Earth because of the loss of the canopy of water that caused the Flood.

Review: Trauma, such as natural catastrophic events, accelerate the aging process. The evidence

demonstrates that one should not take the leap of faith and believe that the rate of decay has always been constant.

The second problem with radioisotope dating is the premise that elements always start from the heaviest radioactive element (the heaviest natural element, uranium), and as they emit radioactive particles over time to become more stable, they eventually become lead. Well not so fast—the theory is partially correct; it is true that unstable radioactive elements decay by losing electrons to become more stable. The problem is that elements don't always start with the heaviest element on the period table. When granite rock (the foundation rock of the crust) is cracked open and samples are analyzed under a microscope, one can see halo rings that are remnants of primordial (original condition) polonium (www.Halos.com by Dr. Robert Gentry). These halos only form when rock cools from a semi-liquid state to a solid state. During this solidification process (which geological evolutionists theorize took millions of years as the planet cooled), radioactive elements emit particles as effervescent bubbles similar to Alka-Seltzer in water. The problem with that is there are ample halos found in granite rock around the globe with polonium that doesn't have rings from a heavier parental element, implying that they began as polonium rather than coming from other elements, such as uranium to radon to plutonium to polonium. Each halo represents when alpha particles are emitted off, and each one leaves a ring. These emitted radioactive particles are fleeting, lasting one to three seconds; therefore, the fact that granite rock has captured these fleeting effervescent rings is irrefutable evidence that the granite rock cooled almost instantaneously. This evidence supports a literal interpretation of the Genesis creation account. If granite rock cooled over millions of years there wouldn't be any captured polonium halos. Dr. Robert Gentry likens this to showing someone a 1,000 years from now the effervescent bubbles of Alka-Seltzer 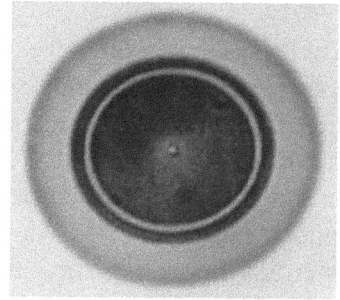 in water; you would have to instantly freeze the water to show the bubbles. The same principle applies with the spherical halos (they are bubble-like) from the emitted particles. The granite rock had to instantly solidify to capture the spherical halos. The image is of primordial polonium halo in granite rock. *Photo credit: www.Halos.com, by Dr. Robert Gentry.*

In addition, the observable evidence that halos of polonium exist in their primordial state means that without the tell-tale rings of heavier elements, polonium was created as polonium and did not evolve into polonium from uranium. The fact that polonium halos exist in a primordial state in granite rock is evidence that they were created as such and did not evolve into such.

Since polonium halos exist without heavier parental halos, they show that the radioisotope dating system is flawed. That particular dating system presupposes that the rate of decay has always been constant. Well, since polonium halos exist without heavier ancestral radioactive elements, the rate of decay hasn't been constant. During creation, the rate of decay process that we perceive today was altered. Similarly to Adam and Eve being formed as fully mature adults without an umbilical cord, the natural aging process and the normal means of offspring were altered during creation.

There has been vehement criticism from the evolutionary community against Dr. Robert Gentry. One was that he was fired from his job. Secondly, critics say that he is a physicists and not a geologist, so he shouldn't be commenting outside of his field. This is an attack on the person, rather than addressing the evidence. Thirdly, the preponderance of evidence from the scientific community agrees with an old earth model, therefore, a majority rule philosophy. Fourthly, the courts heard Dr. Gentry's evidence and ruled against his evidence, allowing evolution to be taught in schools. The courts do not use absolute morality or absolute truth. The courts utilize relative morality and relative truth, which means, what seems most moral and most truthful today is what determines the law. This is why many court rulings are overturned. So the court ruling does not mean that Dr. Gentry is wrong; it means the law today doesn't agree with his findings. Dr. Robert Gentry has addressed each critique leveled against him, and his evidence is irrefutable.

The six-day creation model has granite formed by the end of the first day of creation, albeit in liquid form because the densely packed universe is applying massive amounts of weight and pressure, and this is generating a lot of heat. But once the universe expands on the second day of creation, then the weight is completely removed, the pressure is significantly reduced and subsequently the temperature is reduced as a byproduct, this is when granite solidified as solid rock. The granite cooled almost instantly from a creation model and captured polonium halos, corroborating Dr. Gentry's findings.

Review: The existence of polonium halos in granite rock and in their primordial (first developed/original condition) state without parental halos are evidence that some polonium did not descend from a heavier parental element. Also, this shows that the matter was created quickly, as in the first day of creation rather than over millions of years. Granite rock solidified quickly, as in the second day of creation rather than for millions of years. The existence of primordial polonium halos shows that the rate of decay was not constant in the past.

Are there catastrophic events in the history of the universe that fulfill the requirements to accelerate the aging process of the entire earth? Yes. The first and second days of creation. Those two days were the most violent days in the history of the universe. The first day violently brought about all the matter, and the second day violently expanded the universe. Also, the global flood of Gen. 7. The advent of multiple large asteroids passing through the canopy to get to the earth would have resulted in the canopy coming down (via gravity and covalent bonding of water molecules) upon the earth. And the asteroids impacting the earth would have fractured the crust of the earth, freeing tectonic plates to hydroplane on deep caverns and causing the water hidden under the crust of the earth to burst out, fulfilling Gen. 7. In addition, there would have been many volcanoes that erupted multiple times around the globe, causing massive heat and fires. The beginning portions of the Flood were met with high heat, which is a necessary requirement for the conversion of wood to coal and animals and algae to oil. The fires were put out by the rains, and the high heat was reduced by rain and sun blockage for forty days, which turned the tide of high heat to an eventual glacial age.

The earth has many large impact craters, scars that tell us of a time when multiple large asteroids impacted the earth. So too does our moon tell the same story. It all fits together like one giant puzzle to reveal that the catastrophic event during Noah's day caused all the oil reserves, coal mines, fossils, the seven continents, the glacial age, deserts, continental shelves, decreased oxygen concentrations, increase gravity, and some of the salt mines.

Why would God create the earth and everything in it so that it would appear old? Why would God create an environment so that traumas, such as volcanoes, asteroid impacts, a global flood, and the like, would accelerate the aging process and give the indication that the earth is billions of years old? Why would He then indicate that life has existed for hundreds of millions of years by association? These presumptions would indicate to humanity that the Bible has errors in it regarding time lines and draw into question the Genesis' creation account and the existence of God. Why would God do this since He knows all things even before all things occur?

There are two ways to approach this answer:
1. God wants mankind to believe in Him by faith alone. God wants to know that we love Him for who He is and give Him glory through faith alone.
2. An empirical answer is that God didn't create Earth and everything in it to appear old with a fake history, but He did create everything fully developed; that is, tall trees and Adam and Eve, were fully

developed. Therefore, it's people's desire to prove there is no God by interpreting what they observe to be in contradiction with the Bible.

Let's address the first answer. Before you snicker at the simplicity of the answer, ask the question, "Are we so different?" We were created in His physical image and in His likeness, with His characteristics, thoughts, mind, emotions, and so on (Gen. 1:27). For example, take a man who is super wealthy and is seeking a bride. If that man shows up on the first date with a gold crown, furs, an expensive car, an entourage of servants, and so forth, that guy will never know if the prospective bride loves him for who he is or whether she loves him for what she can get out of him.

How did Jesus, the King of Kings, show up to court humanity—as a wealthy God with processions, pageantry, trumpets, feasts, and a who's who list of guest and attendees? No. He was born in a manger (feeding trough) inside a barn. Later, Jesus the King of Kings and Lord of Lords, said He didn't even have a place to lay His head. He was dirt poor. What about His beauty? Did Jesus come with blemish-free skin. Did he glow or sparkle and have beautiful eyes and strong muscles? Was He taller than others? Was He faster, stronger, or more handsome? The answer is NO, according to Isaiah 53:2: **"He has no stately form or majesty that we should look upon Him, Nor appearance that we should be attracted to Him."** The point is that Jesus wanted us to love Him for who He is, Not for what He looked like or How wealthy He is (He owns everything).

This similar point is seen in many locations in the Bible. And Jesus, though born of a virgin, was asked by the Pharisees, "Aren't you Jesus, son of Joseph?" Jesus knew that the Pharisees were looking for a descendant of King David born of a virgin and fathered by God. He replied, "Yes." Why didn't He say, "Joseph is my legal father, but I am born of a virgin from God"? He didn't want to prove that He was born of a virgin and that Joseph was His earthly father but not His biological father because He wanted to be loved for who He was/is through faith.

Through all the amazing miracles that Jesus performed, each human still has to confess Him as Lord through faith alone. This is the foundation of the Bible: Jesus is the only way to the Father (John 14:6). There has never been a change in the method of salvation from Adam and Eve to all the Old Testament saints to those who saw Jesus alive and to us today. Salvation has always and will always be through Jesus to the Father and no other name.

Every single saved Israelite and gentile, as well as Abraham, Isaac, Jacob, and so on, worshiped and were saved through Jesus, even in the Old Testament. They just didn't know Him by the name Jesus. They knew Him by other names, such as Yahweh, Elohim, I AM, Lord, King of Israel, and others. How did the Old Testament saints worship Jesus? No one has seen God the Father at anytime— correct? Yes. Exodus 33:20: "You cannot see My face, for no man can see Me and live!" John 1:18 and I John 4:12: "No one has seen God at any time." There are other versus as well, but you get the gist. Then all the Old Testament physical appearances of God were not God the Father, but God the Son, Jesus. There are many, many physical appearances of God, and none of them are the Father. Why? For one, God the Father is invisible (Colossians 1:15). Secondly, God became flesh and dwelt among us as Immanuel, or Jesus (John 1:1–14 and Matthew 1:23). Thirdly, Jesus is the Word (John 1:1), and Moses wrote about Jesus (John 5:46), and all the prophets wrote about Jesus (John 5:39). Fourthly, passages from the first verse (Gen. 1:1) ("Elohim" is the plural form of God, meaning Father, Son, and Holy Spirit) to the last verse (Malachi 4:5–6) of the Old Testament/Torah refer to Jesus. And passages from the first verse (Matt 1:1) to the last verse (Rev. 22:21) of the New Testament refer to Jesus. This is why Jesus is called the Alpha and the Omega, the Beginning and the End, the I AM, and the Almighty (John 8:58 and Rev. 1:8). Fifthly, Jesus' own testimony was that He claimed to walk and talk with Abraham, and before Abraham was born, Jesus was the I AM (John 8:49–59).

The second approach to answering why God seemingly created everything to appear old is based on empirical evidence. God didn't create everything to appear billions of years old. It's only mankind's interpretation of the observable evidence that is purposefully directed away from the Bible and away from God. For example, the layers of the crust couldn't be from millions of years of soil deposits because erosion would have commingled the layers, and since leaves decay to dust in months and dead creatures decay to dust in years, therefore, the layers of the crust formed quickly via the Flood to preserve each fossil before they decayed to dust. Earlier, we showed how the processes of making oil and coal and petrified wood don't take millions of years, but days to decades. And all life forms on earth having similar DNA represent one maker utilizing the same dirt and elements to create life, not one original prokaryote as the ancestor. The similarity of the DNA of creatures doesn't mean that we all descended from one ancestor; there is an equally valid interpretation that there was a common designer. For example, look at books written by different authors; they may use the same letters and same words, but that doesn't mean the books are related; they just have a similar designer—humans.

Review: From Adam and Eve to Abraham, Isaac, Jacob, King David, King Nebuchadnezzar, Mary and Joseph, and to you, salvation has always been through faith alone and through Jesus alone. God created earth and everything in it for us to accept Him by faith alone. God didn't create the earth to appear billions of years old; it's man's interpretation that spins the empirical evidence to say it's old.

God spoke twice and wrote twice (in stone to Moses), "I am the LORD your God, you will honor the Sabbath Day, six days you shall work and the seventh day you shall rest, For I created everything in six days and rested the seventh day" (paraphrased from Exodus 20:8–11 and 31:17). A rogue few will attempt, consciously or subconsciously, to get away from the Word of God at any cost. The vast majority of scientists want the truth and are honorable, but when an honorable evolutionary scientist views such a seemingly sophisticated reliable age-testing device as radioactive dating, and the numbers come out millions or billions of years of age, what is the honorable scientist supposed to do? Their eyes are not lying to them; therefore, by default, the Bible's time line is in error, and they accept what the measuring device is telling them. The problem is not science, and the problem is not the honorable evolutionary scientist, but it is with faulty measuring devices that determine age and with those who wish to push an agenda at the cost of truth (this is found in everything a human touches, even religion).
You know what God says about those who deceive with a faulty scale?
Proverbs 11:1: "A false balance is an abomination to the LORD, But a just weight is His delight."
Proverbs 20:10: "Differing weights and differing measures, Both of them are abominable to the LORD."
The faulty radioisotopic dating technique utilizes an inaccurate CRD and is a faulty scale to judge time. What does this mean? It means that if people knowingly use a faulty measuring device to steal, even if the theft is truth, then they are committing acts of abomination against God, and severe judgment is coming upon them. God will visit their iniquities to the third and fourth generation of their heirs (Exodus 20:5–6). Those that unknowingly use a faulty measuring device, such as this faulty CRD, are stealing the truth from the masses, and they are unwittingly persecuting Jesus and being used by Satan to further Satan's agenda. Radioisotopic dating, with its CRD, is the false balance, the differing weights, and the differing balance. Do not utilize this faulty measuring device.

Summary: Radioactive isotopic dating calculates the age of the earth to be billions of years old because of a flawed multiplier, the CRD. Also, radioactive isotopic dating assumes that each element descended down in an orderly time pattern over millions of years, yet the existence of primordial polonium with no spherical halos from a parental element proves a creative process and not a slow uniformitarian (a tenant for evolutionary) process. We have proven that the CRD, though it appears constant, is not and has not remained constant. The fact that granite rock has

captured polonium halos proves that the rock solidified quickly, as in a creative process, rather than from a slow cool over millions of years. Therefore, the ages of all items tested are vastly younger, and the Bible's Genesis account of creation is still time tested and stands firm as the pillar of truth.

Group Discussion:

1. Radioactive Isotopic dating is frequently used and referenced to by the educated and among evolutionists—have you ever had this deceitful measuring device used against your beliefs in the Bible and how did this affect you?

2. Learning that the rates of decay of all things can be accelerated with trauma, how does this shape your understanding of the violence of creation and the Flood?

Chapter 14
Carbon-14 Dating

Carbon dating is not an accurate tool for determining the age of an item. Even evolutionary scientists acknowledge that carbon dating becomes increasingly more inaccurate the older the item is that is being tested. One reason is that our atmosphere is constantly adding radioactive carbon via the sun's energy, and the further back in time, the less carbon14 there was in the atmosphere. When a creature dies, its last breath is a copy of the radioactive carbon concentration in the atmosphere. As time passes, the C14 in the dead creature decreases with a half life of ~5,730 years, and the amount of C14 lost to decay is divided by the rate of decay, and that tells us the age of the item being tested. Well, since carbon dating is comparing the number of radioactive carbon molecules in the item being tested to the number of radioactive carbon molecules in the atmosphere, as C14 increases in the atmosphere from the sun's energy, it gives an older reading of the item's age than reality.

    One of the flaws with carbon dating is that the atmosphere adds 20 pounds of C14 per year as a result of ultraviolet energy from the sun converting nitrogen in the atmosphere into carbon14. That may not seem like a lot, but it is when the whole point of carbon dating is to compare the number of C14 molecules in a dead organic item to the number of C14 (radioactive carbon) molecules in the atmosphere today. For each day that goes by from the death of an organism being tested to the day the sample is taken, the atmosphere is constantly adding C14. Therefore, the process makes the dead organic organism appear older than it really is. When the amount of C14 in the dead thing is reduced by half, that organism is around ~5,730 years old. Do you see the problem? Since the atmosphere is constantly adding C14, then the dead organism will by default have less C14 than the atmosphere and appear older than it actually is. The best way to test this is to compare the amount of C14 in the organism with the amount of C14 in the atmosphere at the time of death, not millennia after the organism has been dead. No scientist knows the C14 concentration of the atmosphere 4,500 years ago. But creationists have an understanding of the amount of C14 in the atmosphere at the origin of life. The answer resides in the canopy of water that surrounded the atmosphere. Water filters high energy from the sun; the more water—the more it blocks. Similar to someone standing in a pool all summer, the body parts that are the deepest have the least tan. Why? ultraviolet rays are reduced proportionately by the amount of water the rays have to pass through. Therefore, since there was a canopy of water that surrounded the atmosphere spherically at creation that morphed into a disk-like formation until the Flood, this blocked the ultraviolet light from penetrating to the atmosphere to convert nitrogen into carbon14. Thus, the further back in time we go, the less C14, but when going back further in time during the existence of the canopy, then exceedingly less C14 exists in the atmosphere. Therefore, when testing anything that existed when the canopy blocked ultraviolet rays from penetrating to the atmosphere, that entity will appear much older than reality.

    There is evidence that the earth once had a canopy of water surrounding it. Currently 50% of the planets in our solar system still have their remnant canopy in the form of rings. Saturn, Jupiter, Neptune, and Uranus have rings.

    The atmosphere adds 20 pounds of C14 per year at the current rate. But we don't know the exact amount the sun's high energy converted N14 into C14 to the atmosphere in past millennia. For example, if an animal had died 5,000 years ago and the sun's ultraviolet and higher solar emissions were blocked from penetrating to earth's surface by a canopy of water surrounding the atmosphere from creation till the Flood (creation estimated at ~4,000 BC till the Flood at ~2,400 BC), so that close to zero C14 would have been added to the atmosphere per year instead of the current 20 pounds per year, then carbon14 dating of any life form near that time period, would be off by a massive amount. Potentially, the dating of the dead organism would indicate that it was 40,000 years older than it really is. And it is a fact that water blocks high energy rays. Thus, C14 dating is inaccurate because there was a watery canopy surrounding the atmosphere that protected life on the earth. Therefore, the description in the Bible of how God created the atmosphere on the second day of creation with water above and water below is the primary reason C14 dating is inaccurate.

Dr. Kent Hovind did a great job codifying some examples of the inaccuracies associated with carbon dating: "living sea mollusks have been dated at 23,000 years old (*Science*, vol. 141, 1963, pg. 634–637). One mammoth had been dated as 15,000 years old at the lower extremity and 45,000 years old at the higher limit. Of course, the operator gathering the data simply threw out the high and low numbers to get a mean average."

There have also been blind studies done. For example, allosaurus bones (allegedly 140 million years old) were sent to the University of Arizona lab for testing to determine the age by the C14 method. The results were closer to the Biblical time line; one sample was dated as 9,890 years old ±60 years, and the other sample was dated as 15,120 years old ±220 years.

Carbon 14 dating cannot be used to date rocks because rocks do not eat, breathe, or drink anything that absorbs C14, such as plants. When someone preaches the age of rocks is verified by C14 dating, they have no idea what they are talking about.

Review: Carbon14 dating is inaccurate and becomes increasingly more inaccurate the older the item is that is being tested because C14 is forming in the atmosphere faster than it is decaying. C14 is virtually non-existent in dead life forms that died before the Flood. The reason is because the canopy shielded the atmosphere from the high energy of the sun, which causes C14 conversion.

Group Discussion:

1. Usually, lay people use Carbon 14 dating to authenticate the very old ages of things. How have their arguments persuaded you against the Bible?

2. Now that you see the limitations of C14 dating, and considering that the earth once had a canopy of water to shield the inhabitants from UV rays, which limited the amount of C14 in the atmosphere before the Flood, how does this shape your understanding of the Genesis creation and Flood?

Chapter 15
Viscosity of Rock

Viscosity is the thickness of something in relation to its motion or fluidity. The lower the viscosity, the more fluid the item is. For example, water has a lower viscosity than molasses, and molasses has a lower viscosity than rock. Everything moves but at different speeds. The speeds vary greatly because they have different viscosity levels. Glaciers seem to be solid and not moving, but with time-lapse photography, we can see glaciers (ice) move as a result of gravity pulling them down slopes. We don't see it live because the viscosity of glaciers is high, so the movement is too slow to detect with the bare eye. We can even change the viscosity of an item. For example, butter seems solid in the refrigerator. But put it in the microwave, and watch butter decrease its viscosity and liquify as it heats up.

Increasing the temperature of an object lowers the viscosity by exciting electrons. Also, applying force to an object lowers viscosity. Both actions increase the fluidity.

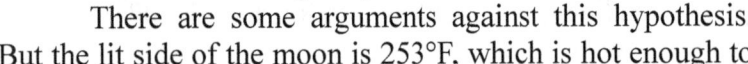

How does this apply to the age of the earth? Well, since all objects are in motion, that means even rocks are moving. However, rocks have such a high viscosity that we do not detect their movements in a decade, or century. Their movements may be detected over several millennia depending on what type of rock they are. The harder the rock, the less movement. So how does this help us in determining the age of the earth? The moon has been hit by many asteroids, meteors, and comets, and those impacts have left craters. All rock moves, and the samples that astronauts have taken of the moon and impact craters tell us that the rock on the moon is basalt rock; it is not a hard rock by Earth standards. If the moon was billions of years old as some suggest, those impact craters would have been flattened out due to the gravity of the moon.

There are some arguments against this hypothesis. But the lit side of the moon is 253°F, which is hot enough to boil water and which lowers the viscosity and increases the fluidity of rock. And the temperature of the dark portions of the moon is –387°F; this change of temperature causes expansion and contraction of the surface rocks. Since the moon's rocks are basalt rock, which is not hard to begin with, then there should be greater movement of rock formations if the Late Heavy Bombardment had hit the moon 4 billion years ago. Therefore, the moon's craters should be smoother and less jagged, but since they are well defined, this suggests recent (in geological terms) impacts within 5,000 years. *Photo credit: http://www.math.nyu.edu.*

Review: The well-defined rocks and impact craters on the moon indicate recent impacts, not ones that are billions of years old. Due to the viscosity of moon rock, the impact craters would have flattened out if they had occurred greater than 10,000 years ago. Ergo, impact craters tell us that every crater we see is relatively recent (less than 10,000 years old).

Group Discussion:

1. How do you suppose temperature swings of 640°F monthly, spanning 4 billion years, should have adversely affected the integrity of the relatively fragile basalt rock and crater formations on the moon?

Chapter 16
Moon Dust

It is estimated that ~40,000 tons of space dust lands on the earth each year. Since the moon is near the earth and travels with the earth, the moon should have similar amounts of space dust collected each year proportionate to its volume and gravity.

When NASA was planning their missions to land on the moon in the 1960s, the scientists calculated how much space dust would be on the moon. Based on the moon being billions of years old, multiplied by the average amount of space dust collected each year, the moon should have 10+ feet of space dust on its surface. It should be noted that when NASA was initially planning on going to the moon, they were worried about sinking because of the moon dust, but closer toward launch, NASA was no longer concerned about any sinking. Yet, NASA still utilized a space module's ladder set ~3 feet off the ground and large saucer landing pads to accommodate any anticipated sinking into the surface dust. One scientist working at NASA wrote, "I get a picture, therefore, of the first spaceship, picking out a nice level place for landing purposes, coming in slowly downward tail-first and sinking majestically out of sight" (*Science Digest*, January '59, p. 36, by Isaac Asimov). To prevent the lunar landing module from sinking out of sight into the dust, NASA built it to have huge round disc pads to land with. It's the same principle of walking in the snow with big snow shoes to prevent sinking. *Photo Credit: https://www.hq.nasa.gov/alsj/a11/a11.step.html.*

When the astronauts landed on the moon, there was not enough space dust for the moon to be 100,000 years old. There was only enough space dust on the moon for the moon to be around 6,000 to 10,000 years old.

That's a problem for those who hold to the hypothesis that the moon is 4 billion years old. How do cosmologists solve this dilemma? They disregard the observable evidence by saying, "NASA was no longer concerned about the depth of the space dust closer to launch. And they'll argue that the amount of space dust deposited on the moon is negligible. They hypothesize that the moon was a wandering small planet that collided into Earth recently in geological terms, and Earth's gravity held the impact debris in orbit, allowing the molten space debris to coalesce into the moon. This is speculative and comes with its own host of problems.

It should be noted that when NASA landed on the moon, they sampled meteors and some asteroid rocks. What is interesting about this bit of information is that evolutionary cosmologists contend that those space rocks and fragments came from the late heavy bombardment about four billion years ago. Why talk about asteroids that landed on the moon? Because there was very little dust on those rocks, and no dust piled up on the sides, suggesting that they impacted the moon recently, such as during the Genesis flood, not four billion years ago. Otherwise, with that amount of time,

25

the asteroids would be covered by space dust, but there was not even a quarter of an inch of space dust covering the rocks. If the moon is as old as evolutionists proclaim, then given the amount of space dust landing on the moon per year, the surface should be smoother than it is. The space dust should have smoothed out the ridged crater walls by now. This suggests recent impacts in geological terms and supports the hypothesis of a young moon.

Review: There is a known amount of space dust collected on the moon each year. Given this current rate of collection, there is not enough space dust covering space rocks and on the surface of the moon to support hundreds of millions of years. However, the amount of lunar dust does support an age of less than 10,000 years. *Photo credit: http://www.exogeologyrocks.com/mineral-and-rock-samples/. Photo credit: http://www.abovetopsecret.com/forum/thread789691/pg1.*

Similarly, each planet in the solar system gets space dust added to it. Well, this provides evidence for a young universe scenario. Saturn is much larger than Earth and therefore attracts more space dust to it. Saturn also has a net or filter to collect the space dust, and that is the rings around Saturn. With the Cassini space probe, cosmologists have discovered that Saturn's rings are very brilliantly lit and very reflective. The reason is that most of the rings are made up of ice. Here's the evidence: if Saturn and its rings were old, such as millions of years old, then the ice rings would be covered in dust and not nearly as reflective. Since the ice is so clean, which is demonstrated by its high reflectiveness and clearly shown by the close-up pictures from the space probe, then the ice rings are not millions of years old, but less than 10,000 years of age.

Evolutionary cosmologists utilize the hypothesis that a comet recently collided into one of Saturn's moons and left a remnant debris of clean ice, free of dust. The problem with this hypothesis is that if a comet collided into one of Saturn's moons, then the debris from the moon would also cover the surface of the ice to obscure the reflective capabilities. Additionally, the heat generated from an impact of a comet traveling 43,000 miles per hour would have melted or vaporized the ice and then coalesced with the debris again, similar to the moon impact theory. And then they would have to pretend that Jupiter, Neptune, and Uranus also had comet impacts because they also have rings.

Review: Saturn's rings are clean of space debris, which indicates they have not been in existence for millions of years, but thousands of years. The rings are made of ice and are still free of space debris that would have eventually dimmed the reflective ability of the clean ice. As it stands, the clean ice, free of debris, stands as testimony in favor of a young universe.

Group Discussion:

1. Have you ever been mocked for believing in the Bible's creation account and how did this affect you?

2. The evidence that asteroids on the moon have very little dust on them, or that there was very little dust on the surface of the moon, which one influenced you the most that the moon is much younger than evolutionists proclaim?

Chapter 17
The Magnetic Field

In the early 1970s, Dr. Thomas Barnes noticed a pattern of measurements of the earth's magnetic field spanning the last 150 samples. When the measurements were plotted on a graph, a pattern emerged showing that the earth's magnetic field has weakened over time. Dr. Barnes calculated a half-life of only 1,400 years and concluded that the earth's magnetic field was less than 10,000 years old. He then surmised that the earth would have a parallel age with the magnetic field because the further back in time the stronger the magnetic field, until life is unsustainable past 10,000 years due to heat vaporizing all life. Evolutionary geophysicists hypothesize that the magnetic fields wax and wane over millions of years, so Dr. Barnes conclusion is only catching one phase of waning. *Image Credit:www.unc.edu.*

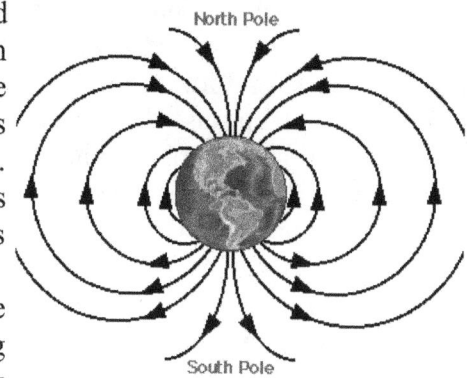

With a catastrophic flood that includes massive movements of tectonic plates, volcanoes, and magma spewing out of the earth, the central core would have had powerful heating because of the global flood. This would have accelerated the decay process of the magnetic field. Considering that the further back in time one goes, the stronger the earth's magnetic field was, and considering the powerful core heating that occurred during the global flood to accelerate the decay process of the magnetic field, then Dr. Humphrey concludes that the magnetic field is about 6,000 years old. And since the magnetic field tends toward disorder and decays over time, then the further back in time, the stronger the magnetic field. This means that we cannot go back too far in time; otherwise, the magnetic field will be too strong and cause intense heat on the surface that would vaporize all life and disintegrate the planet. This supports a young earth with less than 10,000 years of age.

There are magnetic stripes on the ocean floor at the Mid-Atlantic Ridge. The evolutionary geologist hypothesis is that there is a dynamo effect at the center of the earth, and this spinning core slows down to a stop and then starts back up again in reverse, and this causes the poles to flip many times, which caused the magnetic stripes. The problem with this hypothesis is that once the spinning iron core stops to 0 mph, then there needs to be another speculative large force that starts the inner core spinning in the opposite direction. And this process repeats itself frequently. In physics, when a body comes to rest, it will stay at rest unless a directed force is applied. The dynamo hypothesis should be rejected. A better solution that is in accord with Dr. Barnes and Dr. Humphrey, is that during the first day of creation, as matter coalesced toward the core, this increased the density, heat, and pressure of the core so that there was liquid magma at the core. With the core having liquid metal that was rotating, this caused electrons to flow—similarly to a generator that causes copper to rotate and creates electricity with the flow of electrons. As the core increased in heat, it attracted more liquid metal that was charged, this abundance in charged metal and soil generated the magnetic field for the earth, and each celestial body in the universe generated a magnetic field in the same fashion. The stored energy in the core from coalescing matter (first and second days of creation) generated a magnetic field, and this magnetic field, like everything in the universe, tended toward decay and disorder. *Image Credit: www.cmspugliano.wikispaces.com.*

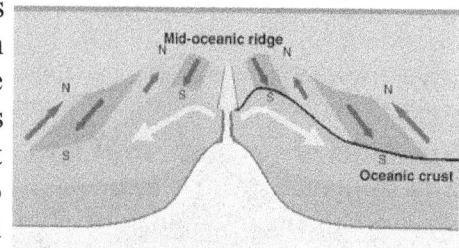

At the Mid-Atlantic Ridge, when viewing the positive and negative polar charge of the igneous rock that comes out of the divergent plate, there is a parallel zebra-like pattern that emerges. As the rock emerged from the crust, it recorded the magnetic field at that instant before solidifying. Thus one may either infer that there have been many magnetic field pole reversals, or that the catastrophic global flood of Genesis caused what appears like 50 pole reversals. The Genesis flood, with multiple asteroids impacting the earth, with hundreds of volcanoes erupting multiple times during the 40-day period of the Flood, with the driving force of the breakup of Pangaea and massively large tectonic plates sliding around the globe on magma as the fountains of the deep burst violently open (Gen. 7:11) would explain what appears as pole reversals of the magnetic field.

However, unlike the sun, that has a magnetic field with many locations of loops bursting out from its crust like twisted coiled springs because the sun has many layers spinning at varying speeds, the earth has a uniform magnetic field emanating from its core. The ability of the sun to flip magnetic poles could be because the core is spinning on its horizontal axis with a slight (9.1° angle per year) vertical rotation as it spins on its horizontal axis. This vertical rotation would cause the poles of the sun to flip every 11 years as heliologist have discerned. One hypothesis to explain the earth's display of magnetic reversals is that the inner core and the crust may have had a slight variation of rotational velocity that caused magnetic pole distortions. Aiding this premise is that the sun is very chaotic and its surface is not rotating in sync with its core, or mantle. This causes its magnetic field to twist and churn. This process of twisting the magnetic field during the Flood saga because the crust was not in sync with its core or mantle, would not form a uniform pole reversals seen on the surface, there would be chaotic locations of where polar magnetic reversals burst out of the crust, just like the sun. Thus, this hypothesis should be rejected.

Dr. Kent Hovind (author of "Creation Seminar" videos), and Dr. Walt Brown (author of "In The Beginning"), both present the most plausible explanation of the polarity reversals at the divergent zones. They contend that there were no pole reversals. Instead, as rock expanded at the Mid-Atlantic Ridge, cracks in the rock filled with cooler water, and the temperature variances formed the alternating polarity in the igneous rock as it diverged out of the crust. Their notion holds merit and should be considered as a viable option to further be explored. *Image Credit: www.mvsdperiod6.wikispaces.com/divergent+boundries.*

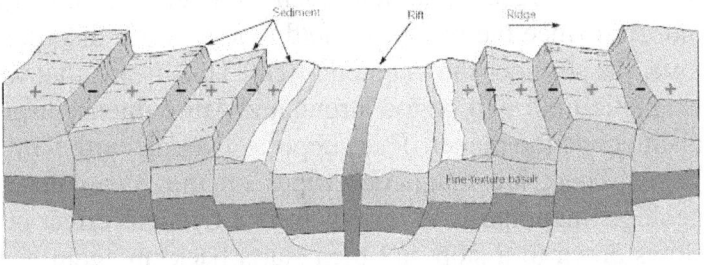

Review: The magnetic field is gradually becoming weaker with a half-life of 1,400 years, which means that as we travel backwards in time, the magnetic field doubles in intensity every 1,400 years. This means that we cannot go back in time further than 10,000 years because the earth's surface would be disintegrated. And considering the heat generated from the global flood, it would have used up magnetic energy and accelerated the decay process, which leads to an age of the earth of 6,000 years. The best theory for the polarity differences on the ocean floor in the igneous rock in the divergent zones, is explained by actions during the Flood with cooler water hitting cracks in the igneous rock, which effected the polarity.

Group Discussion:

1. With evidence that the magnetic field cannot be greater than 10,000 years old, or it would incinerate life, how does this affect your faith in the Bible?

2. Knowing that there are reasonable explanations as to why divergent zones have alternating polarities that are logical and in harmony with science and the Bible, how does that affect your faith in God?

Chapter 18
Polystrata Petrification and Fossilization

Polystrata Petrification: *Poly* means "many." *Strata* means "layers." *Polystrata* means "many layers." Layers of what? Fossils and petrified trees transcending many layers of sedimentary rock.

Let's focus on soil, sediment, and the crust of the earth. The soil of the earth is in layers. Petrification is when a tree decays from wood to stone. We are told that this process allegedly takes 500,000 years by evolutionary geologists. Is that true or just another tale to corroborate the billions of years required for the evolution hypothesis? Science has broken down the mechanics of the petrification process by which wood changes to stone with the following: a tree has to be covered quickly (in geological terms: days to a years) so that the tree doesn't decay to dust. Sediment containing moisture filled with minerals saturates the tree with silica or iron sulfide, and then the tree is baked with high heat and high pressure. *Photo credit: Ian Juby, ianjuby.org.*

The problem is that our top evolutionary scientists tell us that this process takes 500,000 years. And since we know there are petrified trees around the world, then the young earth creationists that adhere to a 6,000-year-old earth have a problem. Right? Both can't be correct. Either it takes 500,000 years to form petrified wood, or it doesn't. Remember, they determine the age of a petrified tree by using a radioactive isotope algorithmic formula:

$$\frac{\text{Pb (lead) ions}}{\text{Po (Polonium) ions}} \quad X \quad CRD = \text{the age of the item}$$

The CRD assumes the decay process has always been constant. Old earth theorists contend that the rate of decay (radioactive and non-radioactive) has uniformly been constant through the ages. The rate of how things age today is the rate of how things aged in times past. This is called the uniformitarian theory. This is a leap of faith because no one was there to observe the rates of aging (decay) in past millenia. And this is one of the reasons that evolution is dependent on faith-based conjecture. There is ample evidence that shows that the rate of decay (aging) can be accelerated with trauma. For example, lightning converts oxygen into nitrogen instantaneously, which causes rainwater to naturally fertilize the soil. Lightning and auroras are examples of how slow radioactive decay rates are accelerated. Both examples demonstrate that the current slow rate of aging (decay) may be accelerated under certain conditions. How do we know the rate of decay is not constant since we can't go back in time to prove it? We can logically conclude this by observing that both nature and mankind can accelerate the aging process via trauma. For example, we can accelerate the decay process of wood to coal in eight months, a process that is supposed to take 20 million years. And we can accelerate the aging process of biomass to petroleum, a process that allegedly takes 50 million years, in 30 minutes. Mount St. Helens' volcanic eruption in 1980 accelerated the aging process of petrification to three decades. And solar flares cause uranium on earth to have monthly and annual variances in the rate it decays. Therefore, what evolutionary scientists are purporting to be an axiom is not even constant in our own generation, let alone for 4.6 billion years. Evolutionists argue that this is an acceleration of the petrification process, not an acceleration of radioisotope decay. The connection is that by utilizing radioactive testing, evolutionists estimate the age through the process of measuring the amount of decay, compared to the rate of decay, from such things as the petrified trees, petroleum, fossils, coal, and so forth. This formula is a modification of the formula physics used for estimating the velocity (V = D/T). But to solve for time, the formula is rewritten to T = D/V, where Distance is how much decay has occurred, with how many lead ions are present divided by polonium. And Velocity is the rate of decay. Any equation with false data will produce false calculations. And the false data in the formula used to determine the age of the item tested is the "Constant Rate of Decay," and there is ample evidence that the rate at which things decay may be altered depending on the conditions. By altering the variables involved in the normal slow rate at which things decay, we may cause those things to accelerate their rate of decaying, establishing that any formula purporting that the rate of decay has been constant is mathematical fiction.

In addition, mankind has been able to accelerate the decay process of petrification in a lab as well. Pacific Northwest National Laboratory has successfully decayed wood to stone in a matter of days. If mankind can petrified wood in a laboratory in days (Science News, February 13, 2005), then there is a problem with this evolutionary hypothesis that it takes millions of years to petrify wood. This establishes that nature and mankind can and have accelerated the aging process. Don't leap to the conclusion that petrified wood is a half million years old because of self-serving evolutionary geologists who merely want to support their preconceived biases. There is a better scientific explanation that is in harmony with the Bible.

Review: Nature and mankind can accelerate the petrification process to decades and days. Therefore, don't believe that it takes 500,000 years to produce petrified wood. The radioisotopic dating formula that determines items to be very old is based on the rate of decay being constant, which it is not. Lightning, high solar energy, chemicals, and trauma, accelerate the aging/decay process.

For wood to be petrified, the wood must be covered quickly in geological terms, within months to years. Why? Wood that is not covered quickly by sediment starts to decay to dust. Arborists teach us that the base of a tree should not be covered up if you want the tree to survive because the bark of the tree will fall off, rot, and decay if it's covered by soil. Therefore, a tree won't wait around 100,000 years for one layer of soil to be deposited. The tree will start and finish the decay process long before one layer is completed. And there are many layers required for petrification. Once the tree has too much soil covering its base, it dies, and the decay process immediately starts. Even petrified trees, when cut open, still have signatures of their annual rings. And the rings usually count in the hundreds, not tens of thousands or millions of years. Thus, further illustrating that petrified trees are covered quickly by soil. *Photo credit: www.geoclassica.com.*

There are hundreds of trees in the vertical standing position around the globe that are transcending through many layers of sediment. Some of them pass through 30 layers of soil. Why is this a problem? Well it's only a problem for old earth theorists. Therefore, when hundreds of tree are found in the vertical, upright position, transcending through many layers of sediment, then that is evidence that the layers were formed quickly, as in less than one year. Sure, some trees are denser and may resist the decay process by decades, but to form a petrified tree or coal, that tree must be covered quickly in geological terms (months to years), before the tree decays to dust. This means that the evolutionary geologist's hypothesis that each layer takes 100,000 to a million years to form contradicts the observable evidence of polystrata petrified trees, not in one isolated location, but around the globe.

A creationist argues against the slow deposit hypothesis for two primary reasons:
1) Erosion would cause rain to mix with the topsoil and would commingle other layers together

as sediment was deposited, preventing globally uniform layers. But since each layer is uniformly smooth, then this represents that either there was no rain on the planet for hundreds of millions of years while each layer of the crust formed uniformly, or the layers formed quickly from a global Flood.

2) When one studies the soil content of the crust, what becomes apparent is that the layers of the crust of the earth are segregated. One layer is limestone, another layer is compressed mud without organic material, another layer is mud with organic matter, another layer is sand, another layer is clay, and so on. This segregation of the layers is not a problem for the Biblical flood, as sediment and biomass mixed in floodwater always settle by their density according to the laws of physics and form uniform and segregated layers. However, evolutionary geologists believe that each layer of the crust formed slowly over 100,000 to millions of years, which is incongruous with how soil is deposited. Essentially, it is impossible to get such exclusivity in terms of the layers because the natural processes of soil deposition does not exclusively deposit only calcium carbonate for a million years, then exclusively deposit sand for a million years, then exclusively deposit mud with no biomass for a million years and then exclusively deposit mud with biomass for a million years. Instead, the natural process is that each type of sediment is deposited at the same time that dead creatures are deposited.

Putting arguments (1) and (2) together, means that the slow deposit hypothesis over hundreds of millions of years is simply faith based and illogical. The Bible records a logical explanation for the existence of the exclusive layers of sediment, and this is the catastrophic global flood. It was not a gentle clear water covering the land; it was a violent judgment with asteroids, volcanoes, fires, ash, debris, acid-rain near volcanoes, fast-moving tectonic plates that forced water out of deep caverns, and so forth. Putting all these pieces together reveals that the floodwater was packed with soil, vegetation, and dead creatures. Young earth creationists contend that the floodwater was not clear and was global, while theistic evolutionists contend the floodwater was potentially clear and regional. A salient point is the fountains of the deep that burst open. A plausible driving force that pushed the water out of the deep caverns as fountains was the fast-moving tectonic plates. With 1,650 years of tectonic plate potentially energy released from asteroid impacts, the plates traveled at 1–2 mph, which forced the waters of the deep to burst violently open; however, theistic evolutionists and "gap theorists" are silent on this point, and both have to reject the words in the Bible that say that every mountain was covered under the heavens (Gen. 7:19–23) and reject the seven-day creation testimony of God (Exodus 20 and 31). They have to use a regional flood for their hypotheses to survive. This makes them guilty of annulling two major events of Scripture. Though a believer cannot lose their salvation—and there are plenty of saved believers that are evolutionists or gap theorists—they could be called least in the Kingdom of Heaven (Matthew 5:19) because they set aside some Scriptures to seemingly fit the Bible with the evolutionary hypothesis of billions of years as the age of the earth, and they teach others to believe the same.

Jesus clearly says that the Creator created man and woman from the beginning, as man and wife. "Have you not read?" Jesus asks in ridicule of people's lack of knowledge (Matt. 19:4–5). God recorded via Jeremiah that evolution is idolatry (Jeremiah 2:27–28). And the Ten Commandments clearly record that God created everything—the heavens, the earth, the seas, and everything in them—in six days and rested on the seventh (Ex. 20:11). And Genesis clearly states that every mountain—all the high mountains everywhere under the heavens—were covered by water. The water prevailed by 22.5 feet (15 cubits) above every mountain on earth and every mountain under the heavens. And all flesh that moved on earth perished—all birds, all cattle, all beast, and all human beings—who were not inside the Ark died (Gen. 7:19–24). There is no way around these two facts recorded in the Bible. Both gap theorists and theistic evolutionists reject these verses as the clear, perfect truth of Scriptures.

Of the two offenders that reject the perfect, sure, right, pure, clean, and true testimony of God (Psalm 19:7–11), evolutionists are the greater offender, as gap theorists at least accept the six days of creation. However, they accept it as a re-creation, not the creation. Both theories have believers, and both theories have those who love God and who are useful for God, yet they have compromised the full extent of their usefulness to appear acceptable to atheists. Every time that mankind has done things its own way, or thought its own way, even though people wanted to worship God, it has always turned out bad. Just ask Abraham about helping God to produce a son his own way; it resulted in Ishmael, the father of Islam, which has been a problem for the descendants of Isaac ever since. And just ask Aaron and the Jews, who decided to worship God in their own way and made a golden calf. This is analogous to evolutionists and gap theorists saying, "I'll worship God my way, and this is through the lens that shows the earth is billions of years old as atheists are telling us." They faithfully reject the accuracy of the Scriptures, which became flesh and dwelt among us.

Listen, the Scripture is clear that we are the representatives of God and should share God with others. This command is very serious; in fact, there is a penalty for refusing to teach others, and there is a penalty for teaching error. Many theistic evolutionists discount the veracity of the Scriptures and just say that it is written by man alone, and thus it has errors, and worse, they teach others this blasphemous teaching. When one views the Bible to have errors, then the result is that there is little desire to read a book that is not supernatural. Thus, there is less knowledge of and obedience to a man-made book. This leads to sin and to stumbling in life. When a theistic evolutionist teaches others the same, they are causing people to stumble. Look at the warning in Matt. 18:6: "Whoever causes one of these little ones who believe in Me to stumble, it would be better for him to have a heavy millstone hung around his neck, and to be drowned in the depth of the sea." That's how severe teaching error is, and that's how important it is to teach truth. Ezekiel continues the point that if you don't teach truth to one in error and they die in their error, then their blood is on your hands:

> Son of man, I have appointed you a watchman to the house of Israel; whenever you hear a word from My mouth, warn them from Me. When I say to the wicked, "You will surely die," and you do not warn him or speak out to warn the wicked from his wicked way that he may live, that wicked man shall die in his iniquity, but his blood I will require at your hand. Yet if you have warned the wicked and he does not turn from his wickedness or from his wicked way, he shall die in his iniquity; but you have delivered yourself (Ezekiel 3:17–19).

If we do not warn them, they will die in their sin, and their blood is required of us for not sharing truth. Why? Life is in the blood (Gen. 9:4–6) and life for a life (Deut. 19:21).

Review: The smooth layers of the soil refute a slow deposit of sediment over millions of years because rain and erosion would have commingled each layer as they were being deposited. Also, the layers are exclusive and segregated deposits. It is beyond rational thinking to believe that only a particular sediment layer was deposited for a million years, and then a different type of sediment layer was deposited for the next million years, and so on. The Genesis Flood produced the smooth and segregated layers of the crust because all sediment and biomass settle according to their density when mixed with water. *Photo credit: In The Beginning, Dr. Walt Brown.*

There are also fossils that transcend through more than one layer. This would be impossible if the layers formed slowly over 100,000+ years. No organism would wait for more than a few months before decaying after death, let alone for 100,000+ years.

One evolutionist that I was talking to about a fossilized fish transcending multiple layers suggested that maybe a fish fell down a hole, then the hole got covered up, and that's how the fossil transcended multiple layers. More savvy evolutionists have argued, in regards to how petrified trees transcend through multiple layers of soil, that a couple of trees accidentally in the upright position transcending through multiple layers means nothing because they probably got disrupted from a mud flow. They say that volcanic mud flows from Mount St. Helens moved buried petrified trees from previous eruptions via lahars (mud flows) into the vertical position transcending multiple layers of soil. This is a plausible argument, but this also is an argument supporting the young earth creationist view regarding how layers of soil can be quickly formed from a flood. Which prior eruption did the mud flow reveal petrified trees from? It would be speculative as well to hypothesize which prior eruption caused the petrified trees to move. Generally, there aren't just a couple of petrified trees in the vertical position transcending through multiple layers; there are thousands around the globe, and they are not in one region near a mud flow—there are petrified trees aligned vertically through multiple layers globally.

Just take a bird's-eye view for a moment. Imagine that thousands of trees on multiple continents are in the vertical standing position transcending through 2 to 30 layers of soil. Can all of those trees be in an upright position because of lahars? That is a leap of faith and not logical to believe. The simple interpretation of the observable evidence is that the layers of soil formed quickly around the trees. This is strong evidence that the soil couldn't have been deposited 100,000 to a million years per layer.

Review: The evolutionary old earth theory suggests that each layer of the earth takes 100,000 to a million years to form; it is proven to be in error with petrified trees transcending through multiple layers. No tree would wait around for one layer to deposit, let alone 30 layers. The tree would have decayed before the first layer was finished depositing. Also, the layers of segregated and well-defined sedimentary layers disprove a slow uniformitarian deposit hypothesis because erosion from rain would have commingled the sediment and soil. Thus, the exclusivity of the layers proves that the layers formed quickly from the global Flood.

The Bible has a record that harmonizes with trees being covered with many layers of sediment in a quick amount of time? Yes. The global flood of Gen. 7. Not only did the water come down upon the earth in rainfall, but water burst up out of the earth. Verse 11 says,

> In the six hundredth year of Noah's life. . . <u>all the fountains of the great deep **burst** open</u>, and the floodgates of the sky were opened. The rain fell upon the earth for forty days and forty nights.

We know that it rained for 40 days and nights, but were the fountains of the deep bursting open for a day, a week, or 40 days as well? The answer comes in Gen. 8:1:

> But God remembered Noah and all the beasts and all the cattle that were with him in the ark; and God caused a wind to pass over the earth, and the water subsided. <u>Also the fountains of the deep and the floodgates of the sky were closed</u>, and the rain from the sky was restrained; and the water receded steadily from the earth.

There is the answer: the fountains of the deep burst open for the same length of time as the floodgates of the sky: 40 days and 40 nights. With water bursting out of the crust of the earth for 40 days and 40 nights continuously, there would be millions of tons of soil being mixed in with the turbulent floodwater. This means the water was filled with soil, vegetation, and creatures—not clear water. A plausible cause of the fountains of the deep bursting open is that the crust fractured from multiple asteroid impacts. The fragmented Pangaea released 1,600+ years of potential energy as fast-moving kinetic energy, and the tectonic plates moved relatively quickly that caused the fountains of the deep to burst open through the fault lines. Fast-moving tectonic plates are accompanied with volcanic activity; thus, hundreds of volcanic activity events occurred during the Flood saga. The fountains of the deep added soil to the waters, as did the volcanoes and asteroids. When the soil settled according to their density over the next year, this resulted in the layers of the crust.

Most theologians believe that the fountains of the deep (Gen.7:11) represent water that burst out of the earth for the global flood, but since there is a great deal of water involved in volcanic activity, then it is possible that the fountains included lava, magma, sediment, and debris that burst out of the deep from volcanic activity. The settling soil that was mixed into the floodwater explains why there are layers on the crust of the earth and explains why there are vertically standing petrified trees transcending through 2 to 30 layers of sediment. The Biblical flood, from start to finish, took almost one year. This quick soil depositing process along with volcanic activity and high heat was necessary for trees to petrify. The observable evidence fits harmoniously with the Bible. The long evolutionary time line of millions of years for layers to form to produce petrified trees does not fit with the observable evidence or with experiments of different soil mixed with water and left to settle.

Review: Multiple polystrata petrified trees transcending 2–20+ layers of sediment have been found around the globe, and they are proof positive that the layers did not form over millions of years but that the layers settled quickly within a year's time.

Group Discussion:

1. How does the knowledge of polystrata petrification and fossilization affect your faith in the Bible?

2. How does Hosea 4:6, "My people are destroyed for lack of knowledge," give you joy as you increase in Knowledge and as your faith is strengthened?

Chapter 19
Distance to the Moon

Distance of the moon to the earth: 238,900 miles. Distance at which the moon cannot be any closer to earth or disintegration occurs: 11,500 miles. This is called the Roche Limit. The difference between the two is 227,400 miles. The current rate that the moon recedes away from the earth is 3.8 cm/ yr.

Some people speculate that the current rate that the moon recedes away from earth has been relatively linear, as they try to determine the age of the earth-moon relationship. They divide 227,400 miles by 1.5 inches (3.8 cm) per year and come up with billions of years. But this is in error. Physicist Donald DeYong explains, "One cannot extrapolate the present 4cm/year separation rate back into history. It has the value today, but was more rapid in the past because of tidal effects. In fact, the separation rate . . . was perhaps 20 meters per year long ago, and the average is 1.2 meters per year." This is a similar error that the radioisotope dating method uses; the rate of decay today, has always been constant. And this is the same error that the uniformitarian hypothesis utilizes; the rate of soil deposit today, is the rate it has always been.

Others speculate what that rate would be using a sliding scale to compensate for larger recession rates of the moon away from the earth in the past and implement a mean (average rate of recession) of 3.94 feet (1.2 meters) per year. Calculating back to the Roche Limit, they come up with 305 million years (227,400 miles / 4.94 ft/ 1 year = 305 million years). This seems like a more accurate process of determining the maximum age of the moon-earth relationship, and it debunks the evolutionary cosmological hypothesis that says the earth-moon relationship started 4.45 billion years ago. *Photo credit: NASA.*

A few new moons are formed from wandering asteroids that get caught in the gravitational pull of a planet, but most moons were formed naturally, as hot matter coalesced to form an orbiting celestial body caught around a larger planet's gravitational pull. This process of coalescing looks similar to the formation of a hurricane. Just like a hurricane that coalesces moisture as it rotates, a coalescing celestial body—during the burgeoning universe—gathered hot matter. At the core of a hurricane is a hole, but at the core of a coalescing moon, planet, or star, would be molten matter. Of the two ways to form a moon, the latter was the prevailing consensus for the formation of our moon until NASA landed on the moon and studied the soil. They observed evidence that the moon had been super hot at one time and that there is a similar isotope with earth. A creationist says, "Well, of course they should have a similar isotope; they were both made equal distance to the sun, and from similar material at creation." An evolutionary cosmologist says, "They have a similar isotope because there was an impact that caused the commingling of the two impacting bodies and shared debris." Cosmologists hypothesize that the reason we don't see evidence of the impact of the moon hitting the earth is because it was completely disintegrated and then coalesced again over time from debris orbiting the earth. And for some reason, it didn't form a ring around Earth like Saturn's, and it didn't leave any distant leftover matter as a ring around the moon, or any orbiting matter outside the moon's orbit as a calling card of the impact. Evolutionists argue that Saturn's rings are a result of a comet hitting a moon. But the impact formed rings; it did not coalesce into another larger moon as our moon supposedly did. These are similar impact theories about Saturn and Earth; one scenario forms rings around Saturn, and one forms another spherical orbiting body called our moon. Saturn's rings are filled with ice that doesn't have dust covering the bright reflective sheen; thus, the impact theory that formed Saturn's rings couldn't include another moon that included dirt. Therefore, it is not a likely scenario to have two moons collide and both contain only water.

However, the most damaging evidence against the lunar impact hypothesis is the lack of evidence. We would suspect to find some matter that didn't coalesce back into the moon, potentially orbiting around the moon, or around the earth, or just beyond both bodies. But as it is, 100% of the matter that got ejected into space got absorbed—this is faith based.

Anyone who believes in the Bible should see that the lunar impact hypothesis is self-serving to promote enough time for evolution and is contrary to the Genesis creation account. For at creation, the earth was formed from water, with an atmosphere and vegetation before the moon was finished forming, and the earth was not liquid magma from the third day on. The moon would have been hot at this point, so there's no problem there, but the earth could not have had an impact with the moon and still have sustained the water, atmosphere, and vegetation. Theistic evolutionists must say that the Bible is wrong or that the evolutionary impact theory is wrong. They are mutually exclusive concepts, and either one is wrong and one is right or both are wrong, but both cannot be correct.

Let's get back to the discussion of the moon receding from the earth for a duration not exceeding 305 million years ago. There is still information missing from the calculation, as starting the calculation at 11,500 miles to not violate the Roche limit, which would cause tidal surges too high to sustain life. Considering that the closer the moon is to the earth, the larger the ocean tides become, then there is a point where life on earth would not be sustainable, because the ocean tides would drown out life trying to take root on dry land and would suffocate life that needs water to survive. For example, at the Bay of Fundy, the ocean tide is extremely large. The tide rises 40 feet and then falls 40 feet. Above the peak tide line, there is ample life, trees, plants, and the like on dry land, but within the tidal area, where there is this chaotic back and forth of the tide, there is less life, a few crabs, sand dollars, and algae. This demonstrates how too much of a change of tide is not conducive for life to thrive.

The distance at which the ocean tides are too high is before the Roche Limit. A safe moon-earth relationship distance starts outside the distance of the Roche Limit and outside the distance at which ocean tides become too large. Going back any further in time and the moon would be too close to the earth and disintegration would start to occur. To actually put a date on when the moon-earth, relationship started and declare the date scientific would be speculative. Whether it's based on the current rate of recession and working backward or on a calculation using a sliding scale with gradually increased recession rates, the conclusions are at best conjecture, since they are not verifiable.

What causes the moon's recession away from Earth? The moon's gravity causes the ocean tides, the ocean tides take angular momentum out of the spin of the earth, and the moon uses that energy to move away from the earth. We don't know what the exact rate of recession was in the past, but the list below exemplifies that the further back in time, the greater the recession rate.

Lunar recession rate per year	Year
3.8 cm	Current
8 cm	1000 AD
24 cm	0
100 cm	1000 BC
500 cm	2000 BC
3 meters	3000 BC
21 meters	4000 BC

Review: Based on the Roche limit, the maximum date that the moon began its relationship with earth is 305 million years ago, not billions of years ago. However, the limiting factor is not the Roche limit for disintegration purposes, but the greater distance required between the earth-moon relationship where tidal forces would not disrupt life. Thus, the lunar tidal forces suggest a younger earth-moon relationship.

Is this a problem for young earth creationists? No, because the starting location where the moon was coalescing matter on the first day of creation was not the place where God placed the moon on the fourth day. During the first day, the moon was close enough in proximity to earth that the outer coalescing material were probably in contact (this helps explain any similar isotopes). During the Big Bang/expansion on the second day, the distance between the earth and the moon was violently expanded away from the earth. And on the fourth day of creation, God finished coalescing the moon, sun, and stars. Therefore, the moon had a rapid recession rate on the second day of creation, which has been gradually slowing ever since. And thus, the distance of the moon to the earth, has natural explanations that are in harmony with a literal interpretation of the creation.

The bigger problem is for evolutionists in explaining how the moon first began its resonance with earth. Their best theory is that a wondering body (the moon) was traveling through space and got caught in the earth's gravitational pull. The two bodies collided, and the moon has been caught by the earth's gravity ever since. This is a poor hypothesis because there is no residual matter left floating in space as debris that didn't coalesce into the moon. It is incredulous that 100% of the floating debris coalesced after the impact. With an impact hypothesis, one would suspect to find some matter that escaped to a greater orbital altitude from earth and remained separate as a ring similar to Saturn's rings. And there is no lunar sediment on earth from the impact either. In addition, the moon hit the earth with a glancing blow such that it did not become one with the earth, yet it was not too much of a glancing blow to cause the moon to float off into space. But the moon does not have enough spin to demonstrate a glancing blow impact. The same side of the moon always faces the earth.

Evolutionists address these problems by suggesting that when the moon collided with the earth, both entities were hot and in a liquid magma state, and this covered up the impact zone as matter transferred from the impact. It is fine to believe that, but don't call it science. It's faith based to say that the earth got the moon from it colliding into the earth. One of the problems with this hypothesis is that the crust of the moon on the far side is ~30 miles thicker than the side we visualize. Now the evolutionary cosmologists believe that perhaps two moons formed after the impact and then they later joined together as one. This is comical. It is best to accept that while the moon was more active, that because of its closer proximity to Earth, there was more tidal friction, which generated volcanic activity mostly on the side most effected by tidal friction—the near side to Earth. And both sides have impact craters, thus, this activity occurred prior to a bombardment of asteroids and meteors. In addition, the moon contains about ~6 billion tons of water ice, which would be expected to find with each celestial body being formed from water as the Bible implies with "waters" being mentioned in Genesis 1:2, and that the earth was made from water and through water in 2 Peter 3:5.

Review: The lack of a debris field, lack of lunar impact scar, and lack of debris on Earth from an impact coupled with a slow lunar spin put holes in the evolutionary hypothesis of how the moon began its relationship with the earth. But the origin of the moon is not a problem from a Biblical perspective, especially with the new discovery that ice water is on the moon.

Mars has two orbiting moons; one of the moons is called Phobos and is about one-third the size of our moon. Phobos is in a death spiral toward Mars and will eventually crash into Mars. The other moon, called Deimos, is smaller and further away, and its escape velocity is greater than Mars' gravitational pull; thus, Deimos is spiraling away from Mars and will eventually be free of the planet's orbit and will drift off into space to collide with Earth or Jupiter or come to some other fate.

Cosmologists hypothesize that they both may be captured asteroids. This seems a likely scenario, considering their formless terrain looks similar to an asteroid. Since one is spiraling into Mars and one is spiraling away from Mars, they both indicate a relatively short existence on an evolutionary time scale. They could not have been orbiting too long around Mars given the rate of Phobos' spiraling descent toward Mars and given Deimos' rate of recession away from Mars. When one works both trajectories backward, it represents a brief period of time according to an evolutionary time span. And in an evolutionary model, there is no recent catastrophic event in the solar system on a scale that parallels with the capturing of two asteroids in Mars' gravity. From a creationist model, both wanderers fit well with the asteroid bombardment period that initiated the global flood, and they very well could be two captured asteroids from that time period. Or they could be moons from creation some 6,000 years ago just like our moon. Either way, there is a nice fit from a creationist point of view. But for an evolutionists to argue that Mars captured two wandering asteroids some time in the past is questionable because their death spiraling orbit and break away orbit, suggest a younger existence.

Review: The two orbiting bodies around Mars, considering their orbits, suggest a relatively short period of time for the existence for their orbits.

Summary: The evidence—moon's distance from the earth, the lack of an impact mark, the lack of an exchange of sediment between the earth moon, the lack of left lunar material that didn't coalesce into the moon, and the lack of appropriate spin—does not corroborate billions of years of existence for the moon. And the impact hypothesis with a molten, magma state earth, is contrary to the Bible that records that the earth was filled with abundant vegetation, seas, surrounded by an atmosphere, and a canopy, before the moon was finished coalescing on the fourth day. Considering tidal forces and the moon's recession, the only plausible explanation is a young earth-moon relationship.

Group Discussion:

1. Considering that the rate of lunar recession away from the earth was greater the further back in time and that this illustrates a young earth-moon relationship, how does this affect your view of the Bible?

2. The further back in time, the closer the moon was to the earth, and this resulted in greater ocean tides. Therefore, hypothetically going back billions of years in time, the moon would be too close and the ocean tides would be too large. And the greater the disruptive tides are, the harder it is for life to thrive in the shoreline, such as at the Bay of Fundy. Thus, do you agree that this undermines the belief that the earth-moon relationship has existed for billions of years, and makes sense from a creationist perspective? If so, why?

Chapter 20
Tyrannosaurus Rex Soft Tissue Found

Tyrannosaurus rex is estimated to have existed 65 million years ago. However, in 2006, Mary Schweitzer discovered blood vessels and cells inside the T. rex bone at a laboratory at North Carolina State University in Raleigh. This should cause all to reevaluate their hypothesis of when T. rex existed.

The evolutionary community claimed contamination of the find. Why? Well, soft tissue can't survive 65 million years, let alone 10,000 years. This is a problem for the evolutionary community, which suggests that the evidence was contaminated, and that is why a 65-million-year-old T. rex had soft tissue inside its bone. Mary Schweitzer, an evolutionist, flatly denies any contamination.

In ideal conditions, red blood cells last only two years in a dead host. The fossilization and cold of Montana is the preserving factor for the soft tissue found in the T. rex by Mary Schweitzer. But how long can soft tissue survive? A group of scientists (Allentoft et al. 2012) calculated the limit beyond which no DNA is likely to survive under optimal conditions. They analyzed mitochondrial DNA and estimated that with a constant cold condition of -5°C in a protective sterile lab, DNA could last up to 6.83 million years before the DNA would be broken down to one base pair. This is the top estimate and under optimal conditions of slowing decay of DNA. For less durable anatomy, such as blood vessels and cells, as Mary Schweitzer discovered, then the decay rate should be much faster. Under warmer conditions, the decay process would be accelerated. Therefore, since Mary Schweitzer's T. rex was discovered in Montana and not at the North Pole, then the decay process would have been accelerated because of periods of warmer temperatures during the summer months and exposure to bacteria. Therefore, a 65-million-year-old T. rex could not have soft tissue still in its bone. Therefore, with Mary Schweitzer's T. rex still having soft tissue, this represents that the specimen is much much younger than 65 million years and could be as young as 4,500 years old. Simply put, soft tissue doesn't last long.

That's not all they discovered about that T. rex; they discovered she was pregnant, which indicates that life on earth was vibrant and healthy for breeding at that time.

Review: The alleged time of 65 million years ago for when Tyrannosaurus rex roamed the earth has more than a serious hole with the discovery that a T. rex bone was found with soft tissue inside of it. The existence of soft tissue in the bone debunks the time line of when dinosaurs roamed the earth. Soft tissue, such as WBC, RBC, and cells, have a short shelf life in a dead host. The cold and fossilization of the surrounding bone would slow the decay process down, but not even close to 10,000 years—let alone the evolutionary ages of 65 million years. The observable and testable evidence allows for a theory that the dinosaur existed ~4,000 years ago and doesn't allow for a theory that the T. rex existed too much beyond that age. And for that matter, since she was pregnant, then environmental conditions at her death were optimal, and we may infer that her death was catastrophic (i.e. the Flood). This is in harmony with the Bible.

Group Discussion:

1. Since the Bible talks about dinosaurs in Job 40 and 41, and soft tissue was found in Tyrannosaurus rex, which is evidence of a young earth, how does this affect your view of the Bible?

Chapter 21
Layers of the Earth

Evolutionary geologists tell us that each layer of the crust of the earth took 100,000 to a million years for each layer to form. And they say that this proves the earth is billions of years old because of all the layers. At first glance, considering the current rate of soil deposits, this old age of each layer and subsequent old age of the earth seem reasonable. But with careful analysis, there are problems with this hypothesis. Look at the layers of the Grand Canyon for a moment. Study all those uniform, smooth layers, does anything stand out to you?

Have you ever wondered if it really took 100,000 to a million years for each layer to form, then why are all the layers uniform? After all, there would be rainfall and erosion during the hundreds of millions of years that would cause mixing of the sediment. Probably, there would be a lot of rainfall and erosion. After all, 100,000 years is a long long time. It's hard to believe that there was no rainfall for each and every single layer. *Photo credit: www.bible.ca.*

The logic glares in our face that the layers had to be formed quickly in order to avoid erosion and commingling of the soil from rainfall. To believe otherwise is ignoring obvious evidence right in front of our eyes.

There is even more evidence to consider, so we don't have to rely on sound logic alone. We can turn to the laws of physics and chemistry for clarity. It is a law that sediment mixed in water will settle according to its density 100% of the time. The reason is based on the amount of mass per volume (density), multiplied by gravity (acceleration), minus the buoyant force of the medium—water. Since each settling sediment layer had the same net gravity applied to it, then the variable that determines the rate of descending to the bottom of water is based on the density. The more mass in a smaller volume, the greater the density; the greater the density, the faster the rate of descending to the bottom of water. Even different densities of liquids will settle according to the principle and form segregated layers. If many different sediments are mixed into a container of water and then the container is thoroughly shaken, when the container becomes motionless, then all the different sediments will form segregated layers based on their density—with the densest sediments at the bottom and the least dense sediments at the top. Vibrating the container will further segregate the sediments into more exclusive layers by breaking static friction. This process is called liquefaction. During the Biblical flood, the vibration that further segregated the settling sediment was a twice-daily, global, and unimpeded, ocean tide, and earthquakes were associated with the fast-moving tectonic plates.

We are fortunate to have the Grand Canyon to expose the layers and the fallacy that each layer formed over 100,000 to a million years without rainfall commingling the layers. Unfortunately, if one goes to the Grand Canyon and speaks to park officials or reads the national park literature, it proclaims that the Grand Canyon was forged over millions of years by the Colorado River. This is in error as well because there are layers of green algae sediment on the upper level of the canyon that is exposed to the hot desert elements. Well, that algae only grows at the bottom of a freshwater lake. And guess what? The algae deposit pattern is in a saucer pattern that matches what lakes produce. The evidence tells us that since the Grand Canyon is at a high elevation of ±6,700 feet, a wall of the lake broke, and all the water stored in the massive lake rushed out quickly to form the Grand Canyon in weeks, not millions of years. When did this occur? It was probably 5 to 10 years after the Flood. The lake had to exist long enough after the Flood for algae to grow and deposit on the bottom of the lake, but not so long that the deposited layers of soil from the Flood that the lake was resting on would have hardened into rock.

Looking at the drawing, which was created by an evolutionary geologist, it is apparent that there are layers. The soil is segregated into homogeneous layers with homologous patterns, and those homogeneous layers lack erosion marks that are associated with a normal hydrological cycle. The young earth hypothesis explains the layers of the crust. The globe was covered with waters that were filled with debris, and 100% of that debris settled according to density. And this caused earth's crust to have different layers in regions around the globe. Because there were some regions that had turbulence and eddies, then those regions had debris that pooled, such as sand in the area of Florida and some massive dinosaur graveyards. But since all matter settles according to its density, the layers have a general outline of similar deposit history because sand settles in water at the same rate no matter where on the globe it is located. If one region has no biomass floating with the sand and one region has massive amounts of biomass floating within the sand, then their rates of settling and the content settled would be altered. The final piece to the puzzle of sorting the different debris into like kind and forming layers was a global tide that swept twice daily around the planet, breaking static friction and allowing further segregation of the debris to like kind. This process is called liquefaction. Also, there were many earthquakes during the Flood caused by the fast-moving tectonic plates. These earthquakes produced liquefaction.

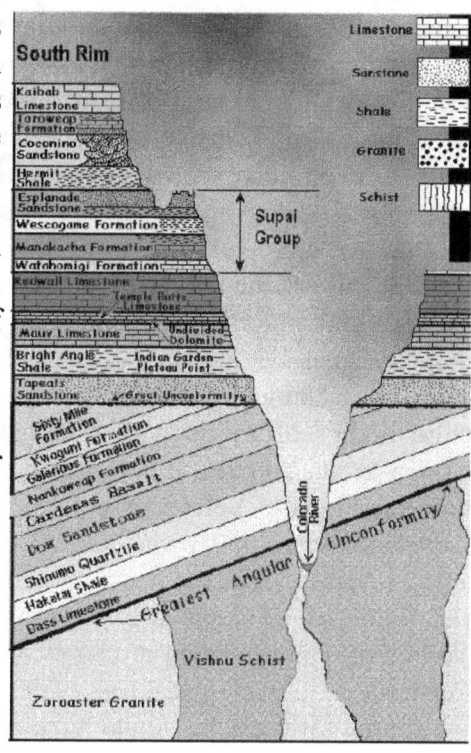

A young earth theorist has the Genesis floodwater filled with sand, mud, clay, silt, trees, plants, grass, ash, and corpses of all the creatures and humans that didn't get on the ark. Therefore, the floodwater was filled with sediment, vegetation, and creatures, NOT clear water at all. The Bible directly says that every mountain was covered by the Flood, every mountain under the heavens, meaning everything. Also, the fountains of the deep burst open violently, as the global flood of Genesis was a violent judgment with asteroids, volcanic activity, and tectonic plates moving fast at ~1–2 miles per hour. *Image credit: The Grand Canyon and the Moon, by Dr. Cowley, Univ. Mich. Univ.*

Review: The layers of the crust are segregated, homogeneous, and without erosion marks from the commingling of sediment that naturally occurs with the hydrological cycle. This fits perfectly with the violent chaotic judgment of the Genesis flood, with the waters not being clear, but filled with lots of different sediment, vegetation, and biomass.

One of the layers of the crust is limestone. Limestone is from broken-down teeth, bone, coral, shells, ash, and the skeletons of plankton made up of calcium carbonate. A creationist has no problem with the evidence of a solid, exclusive layer of calcium carbonate from the many fires and from the living beings that expired during the Flood. The waters roiled with all the sharp debris cutting, pounding, and breaking apart the flesh from the calcium-rich material and formed a layer of bones, teeth, coral, shells, and plankton skeletons equaling many layers of calcium carbonate. Some estimate that at the time of the Flood, there could have been 6 to 10 billion people alive, but that number pales in comparison to the number of land animals that would have contributed teeth and bones to form the limestone layer.

And as abundant as the land animals were before the Flood, the quantity of the mass of teeth and bones adding to the calcium carbonate layer pales in comparison to the tonnage of ash from the volcanic eruptions and fires and also to the tonnage of shells and coral and plankton contributing to the limestone layer.

The evolutionary uniformitarian hypothesis has a problem with this observable evidence of exclusive limestone layers, as they are stuck with explaining away how a solid layer of limestone settled in certain regions of a continent with no other sediment deposited at the same time. Their belief suggests that for a period of roughly one million to hundreds of millions of years, the only deposited soil in several large regions around the globe was calcium carbonate (limestone) and no other sediment. It is illogical to exclusively see a layer of limestone, or a layer of sandstone, or any exclusive layer and listen to evolutionary geologists explain the slow formation of the different layers of soil; for a period of a million years, only sandstone was deposited in a particular region, or only limestone was deposited in a particular region and no other material. For example, in the southern part of the United States, there are limestone deposits hundreds of feet thick spanning several states. And in England, there are massive amounts of chalk, unlike any region on earth. And there are many regions around the globe that have massive amounts of marble. In addition, a slightly acidic body of water would be needed to cause the calcium carbonate and eventually limestone to gather together. They have to find several large bodies of acidic waters to handle the multiple large layers of limestone around the globe. However, acid bodies of water fit perfectly with a creationist's view of the catastrophic global flood, with hundreds of volcanoes erupting around the globe, and the entire saga commenced with several large asteroids. *Photo credit: Creation Faith Facts, by Rob Lester.*

Acid is required in combination with calcium-rich material to form limestone, chalk, and marble. Turbulent waters from a chaotic global flood with hundreds of global volcanoes, which would have been the source for the acid, would have created swirls in the waters that would have gathered together like material with like densities with a similar buoyant force, such as ash, bone, teeth, coral, and shell broken apart by the collisions with other debris in the water. If this pooling of the calcium-rich material was near an erupting volcano that sent enough sulfur matter in the air, then droplets of rain would have combined with the sulfur matter to form sulfuric acid, which would break down the material to form calcium carbonate, or limestone. Evidence that there were massive amounts of heat generated from volcanic activity includes multiple locations around the globe of layers of soil that are bent, with sharp curves. The Gen. 7 violent flood is a perfect explanation of how and why there are sharp curves in rock that didn't break or crack the rocks. The soil that settled in layers was still malleable because of the heat near volcanic activity. As the tectonic plates continued moving, they caused buckling and sharp bends in the soil before it solidified into hard rock.

Evolutionary geologists' explanation of how the crust has an exclusive layer of limestone is that a slightly acidic sea covered an area at some point in time, and coral, shells, teeth, and bone were broken down to their base element over millions of years and deposited at the bottom of the sea to form the limestone without any other sediment being deposited. And then, through tectonic plate movements, the sea shifted and no longer existed over the limestone. But if we took bone, teeth, coral, shell, ash, and plankton and mechanically broke them into smaller pieces because the Genesis flood was violent and soaked those pieces in slightly acidic water, it would not take millions of years to break down the items into calcium carbonate; it would not take 1,000 years; it would not even take 100 years or even one single year. Those items would break down in several months, depending on the Ph of the water. And if a volcano was nearby and enough sulfur was released in the air, then it would precipitate low-Ph acid rain in that general area. The young earth hypothesis of how we got segregated limestone in certain large global regions is in harmony with observable segregated layers, science, and the Bible.

The limestone, chalk, and marble all formed from calcium-rich material. The different end products—limestone, chalk, and marble—depended upon how much heat and pressure was applied to the material. But all three had to have slightly acidic water for the breakdown of the elements to occur. And a catastrophic global flood with hundreds of volcanoes erupting would provide the necessary heat and sulfuric material to convert the waters in certain regions to a slightly acidic state.

And then there are other exclusive layers, such as sand compressed to form sandstone. A creationist has no problem with this—all the sand that was disrupted during the Flood settled and pooled together with turbulent waters and formed layers of sand that was compressed by pressure to form sandstone. However, an evolutionist is stuck believing that for millions of years, only sand was deposited in a large region to form an exclusive sand layer, with no bone, no shells, no coral, no clay, no mud, and no organic material—just sand. Only sand was deposited in regions globally for a million years? This is illogical, yet this hypothesis is ardently defended because the slow deposit theory provides large amounts of time. And time is the foundation for evolution. Remove the large amounts of time, and the evolutionary hypothesis fails.

So how does my hypothesis handle the fact that there are gray, tan, yellow, and white limestone layers, and there are layers of other sediment in between those layers. First off, when ash, teeth, bone, coral, and shell (the constituents that make up calcium carbonate, or limestone) are in their whole form, they each have different densities, though they are made up of the same material. Bone has more air in its trabeculae (tissue) and is less dense than coral. Therefore, the contributors of calcium carbonate in their original whole form each have different densities and settle at different rates. This is one cause of the different layers of limestone. The more the contributing factors are broken into smaller pieces, the more their densities are similar, causing them to settle at similar rates.

Another cause of the different layers are impurities. When a limestone layer forms, it rarely is 100% calcium carbonate. Sometimes there is mud, sometimes a settling layer has some sand in it, sometimes a settling layer has some biomass or adipose in it, and sometimes a settling layer has some vegetation mixed with it. Though the limestone layers are predominately calcium carbonate, there are other soils that are slightly mixed in, and they affect the density and the settling rate. The different impurities cause limestone to be either white, yellow, tan, or gray. And those impurities cause several layers of limestone with different sediment in between them, not because there was one million years of calcium carbonate deposited, then one million years of sand deposited, and so on.

In addition, my hypothesis has many volcanoes globally erupting daily during the global flood, so it is plausible that some portion of the waters and rains were acidic, allowing the broken pieces of bones, teeth, coral, shell, and plankton to form into the base form of calcium carbonate. After the settling concluded, the immense pressure and heat was the final ingredient that determined the type of limestone formation, whether chalk (with the least pressure and heat), limestone, or marble (with the greatest pressure and heat).

When viewing the layers of the crust, another apparent observable piece of evidence is that the layers are smooth with no erosion marks, with no evidence of a hydrological cycle. Evolutionists adhere to a hypothesis that suggests that the layers formed over hundreds of millions of years with no erosion marks, which means that there was no rain during the formation of each layer for hundreds of millions of years. Otherwise, sediment would commingle. A sand layer would erode and intermingle with the layer below, and each layer would have commingling. But this is not what happened based on the observable evidence.

At this point, I'm in awe of the evolutionists' fanatical adherence to their beliefs, while many "Christians" don't even adhere to the Bible with the same fervor. In fact, theistic evolutionists are forced to severely marginalize the accuracy of the Bible to force it to fit their evolutionary views. Evolution does not allow for a global flood, and the Bible directly records that all the mountains and all the earth was covered by the waters in Gen. 7:19: "The water prevailed more and more upon the earth, so that <u>ALL</u> the high mountains <u>EVERYWHERE</u> under the heavens were covered . . . to a depth of 15 cubits [~22.5 feet]." There is no middle ground. Either throw out the Bible or accept the Bible, but don't cherry-pick the Bible just so you don't look foolish to evolutionists.

Continuing this lack of commingling of the layers discussion, there are layers of exclusive clay and layers of compressed mud without organic material (called shale) and other layers of compressed mud with organic material (called black shale). To believe that for millions of years, on a global scale, there was only mud with no organic material deposited on earth and no rain to mix the layers with erosion is illogical. For how could there be no death of animals and vegetation in the million-year period, yet there are dead animal and vegetation above and below that layer of mud? That is incongruous with uniformitarian hypotheses that suggest that there was a slow deposit of soil over millions of years. To have a layer with no biomass means that no life form died and mixed their organic material with that layer of mud deposit for a million years. This is impossible. According to the evolutionary belief in the slow deposit of sediments, there should be no layers at all in the crust, just one giant layer intermingled with limestone, shale, black shale, sand, clay, rock, vegetation, and biomass all mixed together. Evolution is all about randomness. Therefore, having uniform layers and exclusive deposits with millions of years for each layer and no rain to commingle the layers with erosion is incongruous with evolution because it demonstrates order. Something occurred to cause the layers, a very large force that was uncommon and no longer present. However, globally deposited uniform layers are in harmony with quick deposits from a catastrophic global flood.

Review: The uniform, segregated, homogeneous layers around the crust are contrary to a slow-deposit scenario because the natural process is not for each particular sediment to deposit for a million years. This takes too much faith to believe, but evolutionist are forced to abide by this construct to account for the large amounts of time required to evolve life.

Several chapters earlier, we discussed the core samples of the Atlantic Ocean and how there was only ±5,000 years worth of sand deposited on the ocean floor, representing that the Sahara desert has only been in existence for that period of time. And below the tan sand deposited on the Atlantic Ocean floor off the coast of northwest Africa, there is greenish sediment from a tropical forest that existed in the Sahara region before it became a desert. So ponder this, Egyptian soil layers include a layer of limestone from calcium carbonate and a massive layer of sandstone. Keep in mind that sandstone needs massive amounts of pressure to convert the sand into stone. Below that, the sandstone layer is igneous rock. So the slow uniformitarian hypothesis has sand being deposited for multiples of millions of years, given the thickness of sandstone, yet there are only ±5,000 years worth of sand blown into the Atlantic Ocean floor. This means that the observable evidence of ±5,000 years worth of sand existing in the Sahara region doesn't match the slow uniformitarian hypothesis of a process taking hundreds of millions of years. They do not go together. Therefore, either the hypothesis is wrong, or the observable evidence is wrong. The obvious answer is that the observable evidence is not wrong, and therefore the slow deposit theory is wrong. The Bible is in harmony with the evidence.

Review: The observable evidence of when the Sahara Desert formed doesn't match up with millions of years of sand being deposited in northern Africa. Below the sand deposited in the Atlantic Ocean are deposits indicating tropical vegetation, telling us that the Sahara region was

tropical before the Flood, not desolate for millions of years.

Commonly, evolutionists argue that the Gen. 7 flood was regional. As a matter of fact, they have to argue this point tooth and nail, for to give one inch of ground to the Bible being correct means evolution is wrong by default, since they are opposite constructs. If you ever wondered why evolutionists so vehemently argue that they will only allow a regional flood for the Bible legend, it is because there is so much at stake. If the Biblical account of the global flood is correct, then billions of years of soil deposits attributed to the uniformitarian slow deposit theory are lost, and then evolution doesn't have enough time to evolve life. The entire evolutionary hypotheses hangs in the balance. This really is an all or nothing war on truth. The Bible declares the waters covered the earth in Gen. 7:19:

> The water prevailed more and more upon <u>the earth</u>, so that <u>ALL</u> the high mountains <u>EVERYWHERE under the heavens were covered</u>. The water prevailed 15 cubits [22.5 ft.], and the mountains were covered. <u>All flesh that moved on the earth perished</u>, birds and cattle and beasts and every swarming thing that swarms <u>upon the earth</u>, and <u>all mankind</u>; of <u>all that was on the dry land</u>, all in whose nostrils was the breath of the spirit of life, <u>died</u>. Thus <u>He blotted out every living thing</u> that was upon the face of the land, from man to animals to creeping things and to birds of the sky, and they were blotted out <u>from the earth</u>; and <u>only Noah was left, together with those that were with him in the ark</u>.

This is why each mountain has sea shells, and this is why the smooth layers of the crust are global. There is no way around this; it would be logical that it was only a regional flood if there were only regional smooth layers of sediment and if there were only regional mountains covered with sea shells. This would be evidence of a regional flood, but that is not reality. The reality is that there are global smooth layers of sediment, and there are global mountains with sea shells. Psalm 104 reveals that God caused the mountains to rise after the floodwater receded and caused the valleys to sink; therefore, the Genesis flood occurred while the mountains were smaller than today.

A theistic evolutionist once told me that I base my conclusions not on science, but on the Bible. I replied that he was half correct. I don't base my conclusions just on science, but on the evidence, the Bible, and science. I agree that I don't base my interpretations on evolutionary scientists. The Bible is the standard; it is the source of truth and is in harmony with science, not evolutionary scientists. I countered that theistic evolutionists interpret the Bible based on evolutionary conclusions as their standard.

Review: The smooth layers of soil around the globe of the crust of the earth without intermingled sediment proves the long slow deposit of soil required by evolution to be false. There would never be exclusively one type of deposit for a million years, followed by a different type of deposit for a million years, and so on. The smooth layers are evidence that corroborates the global flood record in the Bible, which suggests that the layers formed quickly, as in less than a year. The fact that the layers of sediment are global, and the sea shells on the tops of mountains are also global, corroborates the Bible that the Genesis flood was global, and not regional.

The smooth layers of the crust of the earth reveal that the evolutionary geologists' hypothesis of a slow deposit of soil over hundreds of millions of years is contrary to the observable evidence. And the smooth layers of the crust reveal that the record given in the Bible of the global flood is in harmony with observable evidence.

Additionally, meteorites are absent in the layers of the crust. Since we observe today that around 200 meteorites land on the surface of the earth, in a preserved condition, every year, we would expect to find hundreds of millions of meteorites in the layers of the soil. But since we do not find meteorites in the layers of the soil, we have strong evidence that the layers were formed quickly, such as from a global flood, and not deposited slowly over billions of years.

Does it bother anyone that there are no animal or plant burrow markings in the layers of the crust of the sedimentary rock? If the layers were deposited over millions of years, then there should be some plant roots that burrowed through the layers or some evidence of animal burrowing. But the fact that we see smooth layers without plant and animal burrow markings debunks the slow deposit hypothesis and is in harmony with a quick deposit theory.

From a Biblical perspective, all the ash, soil, vegetation, and creatures that were mixed in the turbulent floodwater from the Genesis flood settled at different rates based on their density over the course of almost one year. The Bible records that after the 40 days of rain and fountains, the waters prevailed over everything for 150 days, then it took another five months for the floodwater to recede into ice (glacial age) and dry up, as mountains rose and valleys sank (Psalm 104:5–9). The duration of the Flood saga is in complete harmony with all the sciences, though not in harmony with evolutionary scientists.

Evolutionary geologists use the different layers of the crust of the earth as proof for the large amount of time necessary for evolution. You see, evolution requires billions of years for random unguided mutations to allegedly have enough time to evolve the genetic code for new functions and new kinds of creatures. Take away billions of years from the layers of the crust, and evolution is broken.

An evolutionary hypothesis of how the many vast salt mines formed is different than my hypothesis that fits a Biblical model. Take for example the Detroit Salt Mine; evolutionists believe that a basin (a hole in the ground) formed 400 million years ago, and ocean water poured into the basin, and then the water evaporated, leaving salt behind. This process kept repeating itself for millions of years until the vast deposits of salt were filled with only salt, spanning 1,500 acres. So no other material deposited along with the salt, no sediment, no vegetation, and no dead creatures? This hypothesis is accepted as fact but is lacking several logical components to the theory, but most notably, the purity of the salt and the lack of other sediment that didn't intermingle with the salt, for instance mud, clay, sand, trees, plants, animals, and so on. It doesn't logically resonate that only salt deposited for millions of years to fill a basin spanning 1,500 acres.

A hypothesis that fits with the observable evidence and science and is in harmony with the Biblical flood is that the glacial age resulting from the Genesis flood caused many of the salt mines (the deeper salt deposits were formed during the high heat of the forming Earth that was surrounded by salt water, the heat separated the salt from the fresh water via a process called evaporation, and this formed very large pure and deep salt deposits, and resulted in the fountains of the deep being fresh water). When salt water has reached the saturation point, any additional salt added remains in a solid form, not diluted with the liquid. When the ocean waters during the Genesis flood started to freeze, only fresh water froze and not salt water, which means that as the glaciers continued to expand, the product left behind that didn't freeze was salt. As the glaciers expanded, the salt quantity increased proportionately. And since salt is denser in solid form than in ocean water, the solid salt sank to the bottom and formed salt mines. This is actually very simple chemistry science, but the reason evolutionary geologists miss this or refuse to accept this is because they are forced to believe in millions of years to preserve the foundation for evolution. But this process didn't take long. It started forming during the time of the waters prevailing over everything for 150 days (this is after the 40 days of rain), and the process continued almost a year. We may deduce the starting point of the ice age by the start of the wind. God started the wind immediately following the 150 days of waters prevailing over the everything (Gen. 8:1). And for wind to occur, there needs to be a large change of temperatures from the freezing polar regions, to the hotter equatorial regions.

Review: The lack of meteorites in the layers of soil debunks a slow deposit hypothesis. People have a choice to believe that salt mines formed over millions of years from oceans pouring into a basin, or the salt mines formed from the global flood, which left behind solid salt. As the glaciers expanded, the salt deposits increased proportionately, and this process took less than a year.

What about the layers of oil reservoirs that are at different depths? The evolutionary hypothesis is that biomass, near a heat source and pressure, was covered by a slow sedimentary deposit over 100,000 to a million years and decayed over a very slow process of millions of years to convert biomass to oil. This, however, is contrary to observable evidence and tests, as we can convert biomass into oil in 30 minutes. And every creature will decay to dust in years, thus, not able to wait around for a millimeter of dust to settle to begin forming a layer of soil. Creationists have the only logical hypothesis, which is that the biomass that lived before the Flood settled at different moments during the Flood. The first and lower oil reservoirs were covered by the initial debris from asteroids impacting the earth, sending massive amounts of soil covering a lot of biomass. Then, as the floodwater rose, other biomass would settle and be covered up by later settling sediment. When the biomass was buried near heat, then petroleum formed from the biomass. When biomass was buried without a heat source, then it fossilized. Oil naturally collects together, so if there were cracks in the settled soil, then the oil would collect. But back to the point of why oil is at different depths; biomass settles at different rates depending on the density of the material, so reservoirs from heavier material would be at a lower level and vice versa. For example, muscle is 3X more dense than adipose (fat), and they would settle differently, forming oil at varying levels. This is one reason there is petroleum at different levels from one global flood.

My concern is that it seems that some evolutionists, even though they know that the natural decay rate of biomass may be accelerated and converted into oil in 30 minutes, reject the above hypothesis, not because it lacks merit, not because it is not plausible, but because it harmonizes with the Bible and debunks billions of years of time. And this precious time is the most important foundation of evolution. Debunking the slow uniformitarian theory, evolution no longer has enough time to account for the diversity and complexity of life.

Review: It's a race against time. Always the creature will decay to dust before enough sediment is deposited to cover it with an evolutionary belief. But with a Genesis Flood, the creature is covered quickly with sediment, heat, and pressure, thus, producing oil, fossils, and layers of soil.

The evolutionary hypothesis is that vegetation, near a heat source and pressure, was covered by a slow sedimentary deposit over 100,000 to a million years ago and decayed over a very slow process of 20 million years to convert trees and plant matter to coal. This, however, is contrary to observable evidence and tests, as we can convert wood into coal in eight months by mimicking the very same process of a buried earth scenario near heat and pressure.

Why believe a hypothesis of a slow sediment deposit when the process requires vegetation to be quickly covered by soil? And why believe a slow deposit hypothesis that suggests that only vegetation was deposited for a period of time and then layers above and below with no vegetation deposited?
Vegetation upon death will not exist long enough for the slow deposit theory to cover it. Decay starts immediately, and most vegetation only last months; some of the sturdier trees may take decades to decay, but that still isn't enough time for the uniformitarian hypothesis. The earth is not filled with a layer of dead animals and vegetation that is waiting to be covered by the slow deposit hypothesis. No, upon death, organisms break down and decay to their base elements. That is why we don't see coal deposits being formed today. It is a race against time—will the dead vegetation decay to its base elements and become nutrients for the soil, or will the slow deposit hypothesis over 100,000 years

cover the vegetation with enough heat and pressure to convert the vegetation into coal? The answer is obvious; vegetation always decays before the soil covers it. In the natural slow deposit hypothesis, there needs to be a quick covering of vegetation with soil, near heat, with enough soil weight to apply lots of pressure to convert vegetation into coal. We are not talking about a thin covering of vegetation; there would need to be lots of soil covering to get enough weight to account for the pressure required.

A hypothesis that is in harmony with observable evidence and science and the Bible is this: The Bible records there was massive vegetation before the Flood, enough vegetation to feed all the abundant herbivores (remember that there were no carnivores until the fall of man). Some of that vegetation got covered up by debris early in the Flood saga by asteroid impacts, tectonic plate movements, volcanic activity, and water bursting out of the earth, sending massive amounts of sediment with it. Those activities sent massive amounts of soil in the air to cover some vegetation immediately, and it was buried near high heat and pressure, and that material was converted to coal in a short amount of time, forming the lowest layers of coal mines. During the middle or later periods in the Flood saga, the vegetation covered by soil either turned into coal if it was near a heat source, or fossilized if it was not near a heat source.

It seems that evolutionists reject the above hypothesis, not because it lacks merit, not because it is implausible, but because they have to reject any hypothesis that debunks millions of years of a slow deposit hypothesis. For evolution's foundation is built on massive amounts of time; destroy that construct, and the entire evolutionary hypothesis crumbles. This is when it seems evolutionist are no longer seeking truth, but are seeking to support their beliefs and destroy the Bible.

Review: The evolution hypothesis of coal formation is that vegetation was covered by sediment over a period of 100,000 to one million years, near a heat source and pressure, and this caused a very slow decay process that converted trees into coal. However, humans can mimic the natural process and convert wood into 100% coal in only eight months. A young earth hypothesis for coal formation is that vegetation was covered quickly by sediment near a heat source and pressure during the varying stages of the global flood saga, and this caused a quick decay process that converted vegetation into coal with different depths. The bottom line is that decaying vegetation will not exist long enough for the slow deposit theory to cover them.

Another piece of evidence in the layers of the crust of earth that debunks the slow deposit conjecture of soil deposited over 100,000 to a million years is the fossil record. The slow deposit of sediment over hundreds of millions of years is one of the backbones of evolutionary theory. The slow deposit hypothesis allegedly creates the large amount of time needed for random, unguided mutations, adaptation and natural selection to evolve life. However, the fossil record debunks this slow deposit hypothesis right off the bat. Evolutionists say that sediment slowly deposits over dead creatures, covering them over hundreds of thousands of years. Once the creature is covered by enough layers of soil for the necessary weight and pressure, then the fossilization process starts—and then that takes 500,000 more years. Could this be true? NO. Think of it is a race with time. It is a race between soil covering the creature and decay. The soil must preserve the creature for fossilization before carnivores, scavengers, bacteria, fungi, sun, and decay either consume the creature or reduce it to its base element as nutrients for soil. It is a race between the decay process of the creature versus the soil deposit rate.

Assume that a creature dies without a burial or a coffin and is just laying down on top of the surface. Immediately, the blood and fluids start to settle because of gravity. Then, within a couple of hours, the body stiffens in a process called rigor mortis. Anaerobic organisms within the digestive tract start to increase, bacteria starts to consume the creature from the inside out, and the body starts to putrefy. The methane gas from the bacteria causes the corpse to bloat. Scavengers arrive quickly to consume the corpse. How long does the slow deposit theory have to deposit enough soil to a) cover the dead creature and b) have enough soil weight to start the fossilizing process before the dead creature is reduced to its base elements, and nothing is left but dust and nutrients for soil?

The answer is that a dead creature immediately starts to decompose upon death and is reduced to fur, teeth, and bones in about two to six months, depending on temperatures and carnivores. And the aroma of a dead corpse attracts carnivore scavengers (coyotes, foxes, lions, etc.) and insects (beetles, flies, etc.) to feed on the dead corpse. With further decay, the stench of a rotting carcass brings the final scavengers, such as vultures, maggots, dung beetles, and more to feast on the remaining flesh. All this is done within weeks of death when the temperature is warm or months when the temperature is colder. At this point, there is still not enough soil deposited to cover a hair of the dead creature. The only thing remaining of the creature is some fur, teeth, and skeletal structure. Then, the final stage of decay is accelerated by the sun. Bones left out in the open with no protection from the sun will start to dry and crack in one year and decay to dust in a couple of years. At this point, there may be one-tenth of a millimeter of sediment deposited, but by this time the dead creature has already been reduced to chemical nutrients for the soil. There is nothing left to fossilize. Therefore, every time a fossil is found in the soil, it is evidence that massive amounts of soil were deposited quickly from a catastrophic event, not from slow deposit over a million years.

Wait a minute. How do we still have mummies then? How do we still have skeletal structures of people buried from 2,000 years ago? Well, the mummies were embalmed to preserve there structures, and all internal visceral tissues were removed to preserve the external tissue. In both examples, the bodies were protected from scavengers and the destructive process of the sun. But most animals are not buried in a memorial service by other animals or other humans. Every time you see a fossil or dinosaur bone that has mineralized into rock, know that that creature or specimen was buried quickly by soil, not buried over 100,000 years. No dead creature would wait around long enough to be covered by one layer of soil—let alone the many layers necessary to generate enough weight for the pressure involved in fossilization.

This doesn't apply to only creatures, but also to leaves, trees, shells, crustaceans, and so on. When a leaf falls to the ground, it immediately starts to dry out and break into unrecognizable fragments from bacteria and fungi feeding on the nutrients of the leaf. No leaf would wait around for years, decades, or centuries for sediment to slowly deposit and cover the leaf. No. The leaf would decay to dust in months. Every leaf fossil is evidence against billions of years, and only allows the conclusion of quick soil deposit covering the leaf. Therefore, every leaf fossil cries out in support of the Biblical flood and testifies against the slow uniformitarian deposit hypothesis.

Look around—there are no fossils being formed today by a slow deposit theory. This should stop everyone in their tracks and pause for a moment. No fossils form today because the rate of soil deposited is way too slow to cover any organism before that organism decays to its base elements and dust. And evolutionists believe that the rate of deposit today is the rate it has always been. There is an obvious disconnect of logic with the evolutionary uniformitarian hypothesis and observable evidence and science. We can observe the rate of decay, test the rate of decay, observe the rate of soil deposit, and test the rate of soil deposit; the bottom line is that the rate of decay far exceeds the rate of soil deposit. Therefore, the only way to get soil deposited faster than the rate of decay is through a catastrophic event. Photo credit: Antiquity: a Review of World *Archaelogy*, edited by Chris Scarre. *www.Antiquity.ac.uk.*

Another problem with the slow uniformitarian hypothesis are the massive amounts of diatomes that have formed the quarries near the San Andrea fault line and other regions around the globe. Diatomes are tiny microscopic organisms in the oceans that sink to the bottom when they die and form a layer of diatamaceous earth. But they only form at a rate of a couple of inches every 1,000 years at the current uniformitarian rate. Finding solid quarries of diatomaceous earth debunks the slow uniformitarian process and is strong evidence that superheated water shot up through the crust in the seas and killed the diatomes instantly. Without other sediment being deposited and mixed together with the diatomaceous earth quarries, as would be deposited with millions of years of slow deposits, they must have been formed by a quick deposit. When superheated water came bursting out of the deep caverns of Gen. 7:11, this would have instantly killed all the diatomes in those regions of eruptions that were covered by seawater and formed the quarries. This theory is in harmony with the Bible and the observable evidence and explains the existence of solid quarries of diatomaceous earth, as opposed to evolution's slow deposit hypothesis.

Also, recall the petrified clams that have been found with their mouths in a closed position. Since the clams are petrified in the closed position, they had to have been covered quickly by soil within days, and died in the closed position because the soil prevented them from opening. Upon the clams petrifying, the mountains rose at the end of the Flood saga and produced the likes of Mount Everest and the Andes Mountains, exposing the clams that were buried alive. Only a Gen. 7 flood explains this observable evidence, and this evidence debunks the slow deposit hypothesis.

Review: Fossils can only be formed when soil is deposited quickly so that it covers the specimen before decay breaks it down to its base chemical elements. Every fossil is testimony that soil was deposited quickly, which debunks the slow evolutionary deposit hypothesis and supports the genesis flood. Diatome deposits accumulate by only inches over 1,000 years, so the existence of solid quarries of diatomaceous earth without other deposits, supports the concept of a quick superheated fountain of water shooting up into the sea. This fits well with the Bible and is contrary to the evolutionary slow deposit hypothesis. Giant clams fossilized in the closed position is testimony that they were covered quickly by soil and pressure, which prevented them from opening upon their death, and thus is evidence for the Biblical flood and against the uniformitarian hypothesis.

There is still one more arrow in the evolutionary quiver regarding the layers of the crust. And the biology department holds on very tightly to this bit of evidence. This is in reference to all the dead animals at different layers in the soil that didn't get converted to petroleum from the heat and fossilized instead. Evolutionary biologists call this the geological column. The geological column is the hypothesis that the deeper one digs down through the layers, the further back in time one goes; through viewing the layers, one can see the evolving of creatures from the primitive lower levels to more evolved creatures toward the surface. And this goes back in time for hundreds of millions of years. When biologists see crustaceans toward the bottom and other seemingly simple life forms at the bottom and seemingly progressively more complex life forms closer to the surface, then this allegedly supports evolution because the billions of years of slow sedimentary deposits laid layers of soil in lockstep with the evolution of life. And they claim that one can see the evolutionary chain of the different life forms as they evolved by looking at different layers of soil and seeing progressively more evolved creatures closer to the surface.

As a side note, don't be alarmed when evolutionists say that Christians use circular reasoning. For example, "Believers claim it's true because the Bible says it's true, and the Bible is the Word of God because the Bible tells us so." This is just a debate tactic to say that Christians don't use the sciences and logic. Though the Bible claims to be truth, anyone can study the Bible and find out what it claimed would happen and what has happened. After careful study, we find indeed that the Bible authenticates itself with fulfilled prophecies and accuracy in regard to doctrine, geography, archeology, biology, and all the sciences. Evolutionists are guilty of the very same thing that they accuse Christians of. They proclaim that one can determine how old a bone is by what level in the soil that the bone came from, and one can determine how old the soil is by what bone comes from that level in the soil. "The rocks do [sic] date the fossils, but the fossils date the rocks more accurately" (Dr. J. O'Rourke, *American Journal of Science*, "Pragmatism versus materialism"). That statement is circular reasoning and not science, but that is one technique evolutionists will employ. They will accuse Christians of the very thing they are committing—circular reasoning.

Why do creatures that didn't get converted into petroleum at different levels in the layers seemingly create a geological column? The answer is twofold. First, there is no place on earth where the geological column is complete. There are a few creatures at different levels in one place on the globe and different creatures at different levels at other places around the globe, and there are no labels on the layers of soil that designates the age. And there are no labels on the bone fossils that designate the age of the specimen. Geologists have broken the layers down to 10 basic strata systems: Cambrian, Ordovician, Silurian, Devonian, Carboniferous, Permian, Triassic, Jurassic, Cretaceous, and Tertiary. But nowhere on earth do all 10 strata systems exist. In fact, Dr. Steven A. Austin, PhD, writes, "Data from continents and ocean basins show that the ten systems are poorly represented on a global scale: approximately 77% of the earth's surface area on land and under the sea has seven or more of the strata systems **missing.**" The only place where the geological column is complete is on the drawing papers from the minds of evolutionary geologists. When creationists say they are skeptical of the geological column that is sternly and dogmatically taught as fact, their objection is based on science. *Photo credit: www.uncommondescent.com/creationism.*

However, even if creationists can prove that a hypothesis of evolution is wrong, it does not mean that creation is correct by default. Since there are creatures at different depths in the soil layers, let's deal with those creatures. The answer to solving the question of why creatures are at different levels of the layers of soil involves bacteria, methane gas from the bacteria, bloating of the host, and subsequent changes in density.

Logistically, creatures that were already living underground before the chaos of the Genesis flood are more likely to stay at the lower levels. Just look around today—we don't see giraffes or elephants or lions living underground; we see simple creatures living underground. And we don't see see crustaceans, worms, and the like living in trees—we see more complex creatures living above ground. That explains why crustaceans and other underground living organism are found at deeper levels because worms and other creeping things were already living underground at the start of the Flood, just like they do today, not because complex life evolved from them while a slow deposit of soil grew along with the evolving life. This is observable and testable.

But what about all the other creatures? How do you explain them dead and buried at different levels, which supposedly supports evolutionary claims? It's actually very simple. When the creatures died, the bacteria inside them didn't die. Oh no, the bacteria had a field day consuming their flesh, with no immune system interfering. And what is the byproduct of bacteria after they eat? It's methane gas. If you ever wondered how methane gas or natural gas got trapped below the crust of the earth and where it came from, this is a contributing factor. The bacteria within each creature were still alive and consuming the creature, though it was dead in the muddy flood waters. And the bacteria consumed internal dead flesh and produced massive amounts of methane gas. But that is not the end of the story. With all the creatures that had bacteria thriving inside them, eating them from the inside out, the bacteria would defecate methane gas. This caused bloating of the host creature, which altered their density. The amount of resistance to bloating from the methane gas varies differently from creature to creature depending on their external structure and their diet. Those with hard shells would resist bloating more than a creature with elastic skin and would remain denser and sink quicker. Those with exoskeletons would mostly resist bloating and remain mostly at their original density. Those with tough reptilian skin would moderately resist and lose some density. Those with a pliable hide but with fur would resist a little and lose more density. Those that had elastic skin with no fur would resist the least, and if they weren't buried by debris, they would float and decay to just bone and teeth in ±six months and further break down to their basic chemical component of calcium carbonate.

If you are struggling with bacteria producing methane gas and causing bloating and altering the densities of creatures to cause them to settle at different intervals, put your forensic hat on and think about a CSI (crime scene investigation) show. Or recall a story on the news in which someone was missing or murdered and their body floated to the surface about three or four days later.

During the global flood, the exterior of a creature was one determining factor in the amount of bloating that altered the creature's density and determined at what level of sediment the creature settled. Another factor is the type and amount of bacteria within the creature that produced methane gas.

Another factor that effected the density of a creature was the ratio of muscle to fat. Muscle is three times denser than fat. A creature that has a lot of fat would be more buoyant than a creature that has no fat. This sheds light on why the geological column has creatures at the bottom that a) already lived below the surface at the start of the Flood, b) had a shell to resist bloating from methane gas, and c) had little to no fat to make them buoyant. For example, top athletes that have very little fat and lots of muscle do not float in water; they sink. Conversely, an obese person with lots of fat and little muscle floats. This is all based on density and buoyancy. *Photo credit: www.genesispark.com.*

When combining these facts about density and things that alter density, then the geological column does not support the evolutionary claim that creatures evolved over billions of years as soil was slowly deposited. That would be contrary to observable evidence and tests; therefore, the geological column demonstrates that living things that died during the global flood settled according to their density along with soil that settled according to its density, and this created the layering effect we observe today and the different levels of different creatures.

During the Flood, the turmoil of the waters caused eddies. These eddies would accumulate animals that ultimately would result in fossil graveyards that are observed in various places. One of the eddies probably formed the whole peninsula of Florida with a large swirling eddy of sand. There is another component to sorting things with the Flood to form the geological column and layers of the crust. When the waters covered every continent and every mountain, there was no physical barrier to stop the tides. The global tide was unimpeded and roamed twice daily across the entire surface of the earth. This led to liquefaction as the tides rolled around the earth, and was the final step of sorting all the soil, vegetation, and organisms according to their density.

When the layers were formed and still malleable from being in warm waters near heated thermal vents, the tectonic plates were still moving relatively quickly, and they bent the malleable layers before they solidified into hard rock. Then the soil hardened into rock with the bent layers. Since the bent rocks are smooth, with no cracking, this indicates that the rocks were bent while they were warm and soft. The Flood solves this, and a slow deposit theory is debunked with this information because hard rock cracks when bent by tectonic movements. *Photo credit: www.allposters.co.*

Review: All organisms have bacteria in them. When hosts died during the Flood, the bacteria still lived inside them and consumed flesh on the inside, and they became bloated from the methane gas. This altered their densities and caused creatures to settle at different rates in different soils. Then upon resting, they either fossilized from the pressure (seemingly forming the geological column) or were completely reduced to base chemical elements or converted to oil if they settled near high heat. The amount of bloating was dependent upon the makeup of creatures' exteriors and the type and amount of bacteria within them. And the final step of sorting the soil, vegetation, and biomass was a large global tide that swept unimpeded across the globe, allowing the settling of sediment and decayed animals and plants according to their densities within the formed layers, seemingly creating the geological column.

A wise person accepts correction and grows in knowledge; a fool rejects correction and mocks the sender of truth and thinks they know everything already. When an evolutionist authenticates an old earth hypothesis by using circular reasoning, such as the age of the soil is determined by the bones and the bones determine the age of the soil, this logic doesn't hold scientific ground and should be rejected.

So where does this leave the evolutionists who believe that the soil accumulated over hundreds of millions of years? Evolution requires billions of years to allow enough time for mutations of the DNA to randomly form new genetic code for new functions and new kinds of creatures. This results in evolutionists defending their theories at all cost, even turning a blind eye to the scientific method and a worthy hypothesis if that construct interferes with the amount of time necessary for evolution. This leads to a faith-based hypothesis and leaves the realm of science.

Remember that the scientific method says that if a hypothesis is proven wrong, then one should start all over with a new hypothesis, yet evolutionists still adhere to soil layers being deposited over hundreds of millions of years and look past the observable evidence.

Review: The smooth layers of the crust of the earth around the globe reveal another violation of evolutionary constructs and reveal the Bible record to again be in harmony with observable evidence. The layers are smooth without commingling from erosion, so this is unmistakable evidence that the soil was deposited quickly, not over hundreds of millions of years.

Group Discussion:

1. Which piece of evidence was the most powerful to authenticate a young Earth?

2. How did this chapter strengthen your faith in the Bible?

Chapter 22
Transitional Fossils

Within the layers of the crust, there are trillions of fossils. They are not rare at all. What is rare and does not exist is a transitional fossil demonstrating a change of kind. Every dead fossil shows that something died, not something that passed on genetic information. Since creatures decay to dust in years and leaves decay to dust in months, then each fossil represents that that item was quickly covered up by many layers of sediment for fossilization and not over millions of years. Evolutionists say they believe in evolution because of the preponderance of evidence in the fossils. But the fossil record does not support evolution at all. What does support evolution are the interpretations of the fossils from their imaginations, the clay models that are created to represent what the archeologist's imagines and the colorful drawings from their imagination. The Bible records the only viable explanation for how we got the layers of the crust with fossils, coal, and oil, and that was from the global Flood. The Flood was filled with sediment, vegetation, and animals, and each settled according to their density.

What about some of those pesky so-called evolutionary missing link discoveries that supposedly prove evolution and debunk creation? They do not exist. They are either falsified findings, or misinterpretations of the evidence, or wrong conclusions because of a lack of knowledge. Keep in mind that though the human body stops growing vertically at a certain age, the facial bones never stop developing. As we age, the jaw protrudes, the eye brow ridges protrude, and so forth. We are able to see these developmental differences from a 20 year old face from a 70 year old face. How much more of a difference do you suppose we could visualize from someone living 900+ years, as in prior to the Flood. And likewise, there would be developmental changes with each creature's skeletal 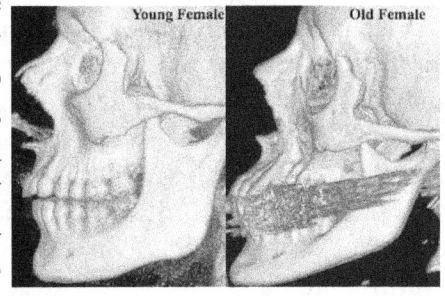 system due to living ten times their current longevity and in a weaker gravity. Over the last 120 years or so, there have been multiple alleged discoveries that turned out to be fabrications, forgeries, errors in interpretations, and so forth. And these hoaxes are on the same level as someone seeing tears coming out of the eyes of a Mary statue. These are all lumped together as hoaxes from fanatical zealots and dishonest people. Let's look at a few falsified claims that set the world on fire for several decades; some textbooks still teach the falsified information. *Photo credit: www.livescience.com.*

Haeckel's embryo drawings: *Image credit: Wikipedia.org.*
German scientist and atheist Ernst Haeckel, who successfully duped the masses in 1860 by falsifying embryo drawings of different species. He drew embryos of eight different creatures and drew them more similarly than they are in reality to make it appear that evolution is true and that we all evolved from similar ancestors. He drew a fish, salamander, tortoise, chick, hog, calf, rabbit, and human. All eight were drawn with nearly the identical heads, midsections (with gill slit folds), and tails. Haeckel drew the embryos in three stages of development. Although his drawings were discovered as a lie, his fraudulent drawings are still in school textbooks as science. Satan, the father of 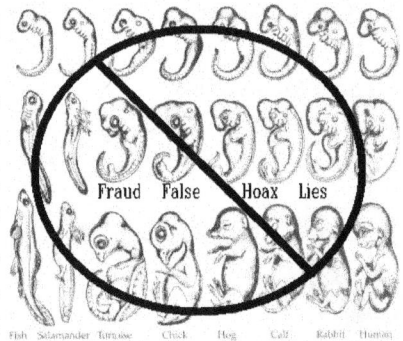 lies, is fiercely indoctrinating the masses of all ages. The folds in the skin of the embryos are not vestigial gill slits that have evolved away; they are folds in the skin that represent different neurological parts of the anatomy. *Image credit: yoursincerefriend.blogspot.com, by Harun Yahya.*
Noted evolutionist Stephen Gould had this to say about Ernst Haeckel's drawing:

Haeckel's forceful . . . books appeared in all major languages, and surely exerted more influence than the works of any other scientist, including Darwin . . . in convincing people throughout the world about . . . evolution . . . Haeckel had exaggerated the similarities [between embryos of different species] by idealizations and omissions. He also, in some case—in a procedure that can only be called fraudulent—simply copied the same figure over and over again . . . Haeckel's drawings, despite their noted inaccuracies, entered into the most impenetrable and permanent of all quasi-scientific literatures[sic]: standard student textbooks of biology.

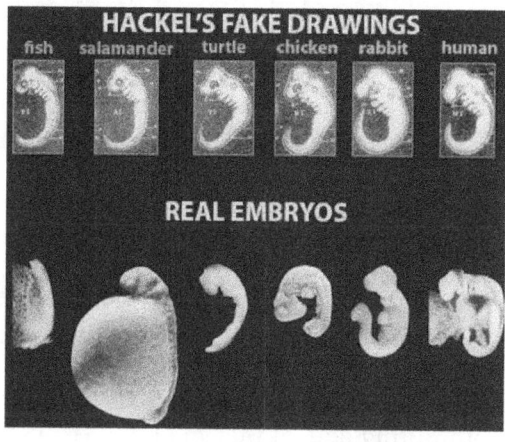

One of the fathers of evolution falsified information to push evolution on the public, and it worked. Imagine the outcry today if Jesus falsified His information to push God the Father. The Bible would be rejected. Combine this with the notion that the early fathers of evolution, including Darwin, erroneously thought the cell to be a simple protoplasm of ooze, and all that was needed for the cell to form was a crystallizing effect of complex chemicals to form the first single-cell amoeba. Darwin believed that a crystallization process was how the first living organism evolved from nonliving material because he didn't understand the cell. They were gravely lacking knowledge. Cells are complex. Imagine if Jesus lacked knowledge about things He professed to be an expert on, and based whole theories on a completely wrong foundation as Darwin, the father of evolution did. Then the entire Bible would be completely rejected.

Also, the early fathers of evolution thought that some species of humans had evolved to a greater degree than other species of humans and that some humans were just barely past evolving from primates, while superior white people had evolved faster than their black human counterparts. Why on earth would any black person accept this lie, and for that matter, why would any God-fearing believer accept a hypothesis that suggests at its origins that one race was inferior to another by some evolutionary process. Imagine if Jesus was a racist as Darwin was; the implication is that the entire constructs of the Bible would be rejected. The Bible clearly states that God is the Father of all and that there is no difference between any race; all races stem from the Tower of Babel (Gen. 10), from which God scattered them around the globe, and 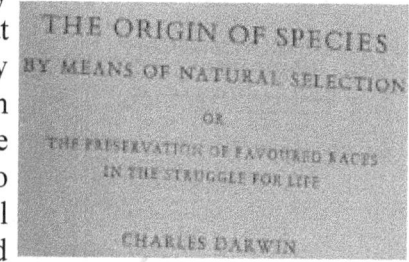 every race is an offspring of Adam and Eve. But this is not so according to racist Thomas Huxley (an evolutionary forefather), who wrote, "No rational man, cognizant of the facts, that the average negro is the equal . . . of the white man." And <u>Charles Darwin, also a racist, was so blinded by his racism that he titled one book, *On the Origin of Species by Means of Natural Selection, or the Preservation of Favoured Races in the Struggle for Life*.</u> Very few people have heard the full title, most only hear it titled "The Origin of Species." Darwin also wrote,

> At some future period . . . the civilized races of man will almost certainly **exterminate** and replace throughout the world the savage races. At the same time the anthropomorphous apes . . . will no doubt be exterminated. The break will be rendered wider, for it will intervene between man in a more civilized state, <u>as we may hope</u>, than the Caucasian, and some ape as low as a baboon, instead of as at present between the negro or Australian and the gorilla.

Charles Darwin was certain that the white man would widen the evolutionary gap by exterminating the black man and the gorilla, so that the evolutionary gap between the white man and the next lower species would widen to the baboon. This list of racist quotes continues with Henry Fairfield (another evolutionary forefather): "The standard of intelligence of the average adult Negro is similar to that of the eleven-year-old-youth <u>of the Homo sapiens</u>." And why would a woman believe in a man-made hypothesis that was fathered by sexist men who said,

> The chief distinction in the intellectual powers of the two sexes is shown by man attaining to a higher eminence, in whatever he takes up, than woman can attain—whether requiring deep thought, reason, or imagination, or merely the use of the senses and hands.

The Bible clearly states that all, including members of both genders, are to submit to everyone (Ephesians 5:21), and both male and female were created in the image of God (Gen. 1:27), and both genders are equally loved by God, and both genders may enter heaven (John 3:16). Imagine if Jesus was a sexist like Darwin, the father of evolution. The implication is that the entire Bible would be rejected.

If someone opposed God the Father, Jesus, or the Bible, were to write a book to discredit them, imagine all the scandals they could cite from those that claimed to follow God. The point is that an attempt to discredit either all honest Christians or all honest evolutionists by pointing out only their most embarrassing moments is not a balanced approach in truth seeking. However, when the ones committing acts of racism and sexism, expressing a lack of knowledge about the complexity of the cell, perpetrating fraud, and making gross misinterpretations happen to be the fathers of the evolutionary hypothesis, and they used those things to further their propaganda on the public, then this is a severe indictment on the premises built by them. Though alleged Christians too have committed acts of immorality, the Bible remains unchanged. It is not the forefather of Christianity (Jesus) committing these acts of immorality; it is the forefathers of evolution committing these sins. Thus, criticism of the racism, sexism, fraud, deceptions, and lack of knowledge is valid and worthy criticism of evolution because the forefathers used those things to establish evolution. The converse against the Bible is equally true; if you find such things against God, then render just criticism against the Bible built on God, if you cannot find such things against God, then be silent against the Bible. *Image credit: Wikipedia.org. Neural development in humans.*

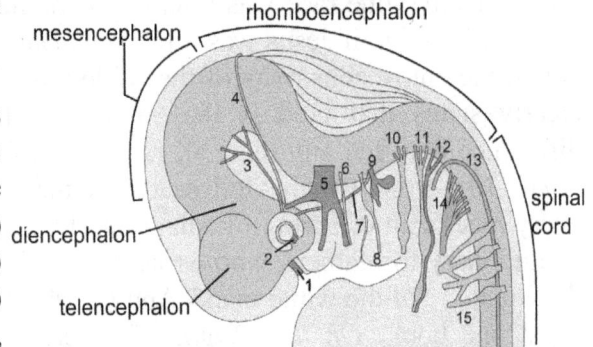

The truth about the folds in the embryo skin is that they have zero to do with gills from a fish; they are a development of the brain and nervous system, with the following functions: (1) olfactory, (2) optic, (3) oculomotor, (4) trochlear, (5) trigeminal sensory, (6) trigeminal motor, (7) abducens, (8) facial, (9) vestibulocochlear, (10) glossopharyngeal, (11) vagus, (12) cranial accessory, (13) spinal accessory, (14) hypoglossal, and (15) cervical n. I,II,III,IV. It is easy to see that the folds have absolutely zero to do with gills.

Don't let evolutionists continue to fool you by citing Haeckel's fraud. Evolutionists may continue to believe the brachial clefts are vestigial gill slits from ancestral fishes, but that is not science; that is a faith-based belief.

Review: Ernst Haeckel falsified his embryo drawings a 100 years ago, and he has duped the world in fast-tracking evolutionary acceptance. The early fathers of evolution were in error and thought a cell was a simple ooze of protoplasm, and life only needed to crystallize into forming

the first self-replicating cell. The early fathers of evolution built the foundation of evolution based on lacking knowledge of the cell, racism, sexism, and forgery. If Jesus had done any one of those things, the Bible would be rejected. But instead of rejecting evolution because the forefather's prejudices and errors, evolutionists have built an ever-larger construct upon them.

Piltdown Man:

A grand scale paleoanthropological hoax that surfaced to force the acceptance of evolution occurred with the alleged discovery of the missing link that bridged the fossil record of when primates evolved into humans. This discovery was called "Piltdown Man" and was found in East Sussex England, 1912. How long did this lie permeate academia? Until 1953—it took 41 years for it to be exposed as a forgery. Apparently, this archeologist took a jawbone of an old orangutan and combined it with a human cranium; the fragments were chemically stained to artificially age them, and the teeth were filed down to add to the ruse. It was a blatant lie. How many weak-minded people threw away their Bibles to accept this tall tale.

One of the notorious names involved in perpetrating the Piltdown man farce was Grafton Smith, the one who trained Dr. Black, who allegedly found "Peking Man."

Nebraska Man:

This was another scheme to artificially advance the cause of evolution and dupe the masses. The archeologist proclaimed to have found the missing evolutionary link in the fossil record that proves mankind evolved from primates. This wile was based on a single tooth—that's right, one single tooth—and from that, a whole theory supporting evolution was extrapolated that man evolved from an ape. And this was the proof. The tooth was examined in 1922. *Photo credit: Wikipedia/Nebraska man.*

Later, in 1926, additional inquiries at the dig site occurred, and another identical tooth was discovered from the jawbone of a pig, and the claim was retracted. But the damage was done. Evolutionists got their headline. How many heard the declaration and never heard the retraction? Often, the cover story is the proclamation, and the retraction is buried in another paper. It makes one wonder how many people threw away their Bibles from this false tale or incorrect interpretation.

How many drawings and full-scale models cleverly created by some imaginative evolutionist to bolster their claim from this one tooth still appear in museums and textbooks to deceive the masses.

Java Man:

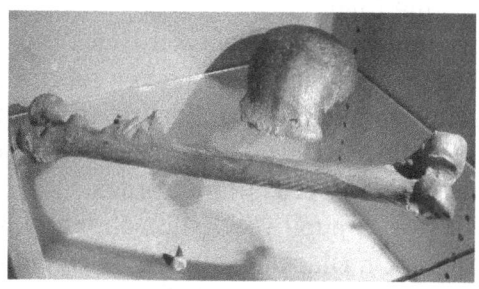

Dutch evolutionist Eugene Dubois set out to find the missing link in the fossil record that connected the evolutionary chain and prove man evolved from primates. This was his goal and his mission; this is what he earnestly believed. This is important to know because he didn't go to Java Island to search out different flowers and accidentally stumble upon this discovery. No, this was premeditated; in fact, he was a pupil of Ernst Haeckel, who was exposed for falsifying his infamous embryo drawings. *Photo credit: Wikipedia/Java man.*

In 1891, Dubois proclaimed that he had found a transitional fossil that proved humans evolved from ape ancestors. Located on the island of Java, Indonesia, he "found" a tooth, a skullcap, and a femur. However, it wasn't Dubois that was at the dig site. He was situated comfortably at headquarters and corresponded via letters and took an occasional horseback ride to the dig site. When workers found bones, they brought those items to Dubois, who was off site, and he made the leap of faith that the bones were from the same dig site and connected each together. The bottom line is that Dubois, with his preconceived bias established, he determined that the bones were the "missing link," ignoring the possibility that the bones were from different dig holes and different layer levels. Compounding the problem, there was no geologist on the site and no objective set of eyes to oversee the operation. He already had his mind made up before he even went to Java as to what he was looking for and would find. Is this a case of Dubois just misinterpreting the evidence or something more sinister? Let's give him the benefit of the doubt and say he got caught up in the fervor of the moment and was so convinced that man evolved from primates that his subjective bias prevailed over his objective conclusions. However, it seems that he concealed evidence to bolster his claim. For example, there were other bones that were also found at the dig site by the workers, such as two skull caps, but he kept them away from public viewing and hidden because they were human, and that didn't fit the evolutionary time line. And he stopped allowing others to examine his Java fossils, which meant it was his word alone, and there was no objective analysis by critics.

Dubois argued that the specimen was the missing link and called it Anthropopithecus erectus, which is a type of specimen from Homo erectus, and declared an estimated 700,000 years of age. His claim set the world on fire with the debate of evolution versus the Bible, and there were almost 80 books written about this "discovery" in less than 10 years time. When the link grew tenuous, and because many knew this was not the missing link from apes to humans, many evolutionists shifted gears and just said that Java man was a primitive side branch of the evolving process or evolutionary tree of life, but they still held tight to their evolutionary views.

The femur was found a full year later about 40–50 feet away from the original skullcap. And Dubois' failure to document the exact soil layer and dig site disqualifies his claim. Many who examined the bones became skeptical of his claim, such as Herbert Wendt, *From Ape to Adam* "this creature was an animal, a giant gibbon, in fact. The thigh bones has not the slightest connection with the skull." Dubois went into seclusion when skepticism rose, and when a German team of 75 researches went to Java Island to corroborate his claim, he refused to help. The team went to the alleged original dig site and unearthed massive amounts of soil and found 43 crates worth of bone fragments that were labeled and sent off to Germany for analysis. A geologist on the expedition refuted Dubois' claim and determined that his Homo erectus claim was a modern human fossil and regular primate fossil combined and not a missing link.

Sadly, today there are evolutionists that cite Java Man as their evidence of proof of evolution. The bottom line is that Java Man is not a missing link or a transitional fossil.

<u>Peking Man:</u>
Evolutionist Dr. Davidson Black did most of the studies in the area until his death in 1934, and then Franz Weidenreich took over shortly after that. The dig site was in Beijing, China. It is important to know something about the person who makes the claim of finding the missing link in the fossil record that allegedly establishes that humans evolved from primates. It is not that Dr. Black was out and about digging for seeds or meteorites and he happened to stumble upon this discovery; no, he had a mission. He was a pupil of Grafton Smith, who was a known evolutionist and involved in the Piltdown Man fraud. Dr. Black already had his mind made up and was looking for bones to prove his beliefs.

The digs first find was a single tooth. After examining the tooth, Dr. Black concluded that it was two million years old and that it had likely come from a new and separate genus of man. He named the alleged new genus Sinanthropus pekinensis, which means "Chinese man from Peking." Really, a separate genus of man, from a single tooth allegedly two million years ago? You can tell his beliefs were dictating his interpretations, just like evolutionists do today. In total, the site yielded 40 specimens and was a success for finding things that had once lived, died, and were covered quickly by soil.

The specimens were called Peking Man, though they had roman numerals to differentiate between them. The dates the specimens were discovered ranged from 1923–40, with the actual Peking Man being found in 1929, which was a skull cap in Beijing, China. A skull cap is the top portion or fragment of a skull. Peking Man was heralded as a missing link, which is a transitional fossil allegedly proving that mankind evolved from primates. As war broke out in WW2, the fossils were lost, and the claims were never verified by an objective team. The bones that were found were classified as Homo erectus, and even though they have been lost for almost 80 years, they have recently (March 2009) been aged at 750,000 years old, based on notes and alleged plaster casts taken of the skull caps just prior to their timely demise. The accuracy of dating something that has been lost since the outbreak of WW2 is highly suspect, and basing them on casts that are not assured to be accurate copies or modified fabrications is highly suspect. The fact is, no one will ever know the truth of what happened to the Peking Man or what it really looked like because the originals are gone, and with only one plaster cast of the specimen, we are not assured of its accuracy or if there was any modification to the cast. But imaginative evolutionists still created full-scale models of Peking Man for museums, and drawings of Peking Man litter the textbooks to further teach the masses about evolution.

Allegedly, the dig site found both human and ape bone fragments—a likely scenario to explain why the ape skulls are broken and fragmented is that ancient Chinese humans hunted the apes for food. Another possibility is that both humans and apes were living relatively near each other, and when the global flood occurred, they were killed by water and debris fragmented their skeletal structures. Then, with the turbulent waters, they settled down together.

Orce Man:
This discovery was made in southern Spain in a town called Orce in 1982. it was proclaimed to be the oldest fossilized human remain ever found in Europe. This skull fragment was said to be a young 17-year-old male from around 900,000 to 1.6 million years ago. Evolutionists created nice drawings of what this person looked like, and of course they made him look like a transitional person halfway between an ape and a human. With the short brow and prolonged slanted jaw, it was very typical evolutionary propaganda stuff.

Now, when they say they found a skull fragment, they literally are talking about a fragment that is 3–4 inches in diameter and almost circular; we are not talking about a full skull or a complete bone of the skull. Extrapolating a story out of a bone fragment is not typical of honest evolutionists, but objectivity is strained to its limits when preconceived biases accompany well-meaning evolutionary archeologist on dig sites. A skull fragment smaller than the palm of a hand and a hypothesis spun the size of Texas comes out of it that supports the preconceived beliefs already entrenched before the excavator even grabs a shovel. And then the unknowing masses go to a museum, and they see a fully developed skull or a fully developed face, or a full-scale model of an alleged Homo habilis or alleged Homo erectus or alleged Neanderthal man, not knowing these fabrications are based off a single tooth or small bone fragment or falsely combined bones from multiple sites from different species to form one transitional species. Let's give the evolutionary archeologists the benefit of the doubt and say their biases overcame their objectivity, rather than saying this is a grand forgery to support a grand tale.

Dr. Duane Gish, "Evolution: The Fossils Still Say No!," has a critical review of the evidence and logic that debunks the Orce man claim and reduces the claim to a young donkey's skull fragment, not human, and exposes the ridiculousness of basing a fictional tale from one skull fragment.

Sahabi fossil:
This is a collarbone discovered by Noel Boaz in the Libyan desert in 1979. Mr. Boaz claimed that this

clavicle belonged to a primitive ape-man who lived around five million years ago. It is claimed as the oldest "ape-man" fossil in the world. He claimed that this ape-man walked upright. All this was from one bone? Again, this was an entire story extrapolated from one bone and a presumptuous proclamation of five million years of age. As it turns out, when objective analysis was performed on the alleged missing link, this "collarbone," was not "S" shaped like normal collar bones; no, it was "C" shaped like a rib and turned out to be the rib of a dolphin.

Neanderthal Man, "The Old Man" of La Chapelle:
The discovery was made in 1908 in La Chapelle-aux-Saints, France. The bones of this specimen included a skull, jaw, most of the vertebra, several ribs, most of the long bones of upper and lower extremities, and some bones of the hands and feet. Now here is the problem: the one who made the discovery wanted to find fossil evidence of evolution, so upon proclaiming to the public what he had found, he declared it to be a primitive creature about 60,000 years old and a transitional stage between primates and modern humans. And when a model was made of what this specimen looked like, they made him look extremely like a transitional being between gorilla and human to support evolution as much as possible. Many people believed this claim and the fabricated model from 1908 until 1957, when an independent review of the fossil occurred.

According to Dr. Rudolf Virchow, a top scientist from Germany who laid the foundation for modern pathology, studied the evidence and determined that "Neanderthal Man" was a feeble Homo sapien (just a human being), who had a head injury that disfigured his skull, deformities, arthritic conditions, and rickets. Neanderthal Man was an old man who had lived a long life. Almost all his teeth had fallen out, and regrowth of the mandible and maxilla had already set in as though there was a decade of eating food without teeth. Additionally, and here is the kicker, this person had very severe osteoarthritis. All the alleged changes in the bone structure that were associated to evolutionary modifications of the skeletal system were merely severe deformities from arthritic changes. Nothing evolutionary at all, but if you go to an evolutionary museum, you will read all the "facts" about the missing links and see full-scale fabrications from a bone fragment. In fact, the ability for the bone to adapt to variances of stimuli is based on encoded DNA (adaptation), not variances in the bone from external stimuli resulting in enhancements in DNA (evolution). It is the preexisting DNA information that results in the body's attempt to stabilize a damaged joint with traction spurs, schlerotic changes, and joint fusing—which are arthritic conditions. The impairment of function does not add functional information to the DNA code for any future new function or new kind of creature as evolution proclaims. And this "Old Man of La Chapelle-aux-Saints" is not a stage in the transition from primates to modern humans, but a human being with a severe arthritic condition, which is adaptation.

Two modern-day scientist, Dr. Straus, an anatomists from John Hopkins University, and Dr. Cave of St. Bartholomew's Hospital Medical College, reexamined these bones of Neanderthal Man in 1957. They concluded that Dr. Rudolf Virchow's finding were correct. Addressing the arthritic condition of the "Old Man" does not give an explanation of any of the other skull caps that may be found that are seemingly near human or that seem to be some Neanderthal creature. Think of it this way: while humans age, their facial bones continue to develop. The Bible records that human beings used to live 900+ years before the Flood. Well, this would have created pronounced eye ridges, jaws, and so forth. The interpretation from the evidence is where things go haywire, preconceived beliefs of evolution dictate the interpretation. Neanderthal Man is not a transitional fossil, though he is still referenced by many evolutionists today.

"Lucy," Australopithecus afarensis:
Lucy was estimated to be 3.2 million years old and was discovered in Ethiopia in 1974. She consisted of an alleged bipedal upright gait. Countless models were fabricated to illustrate this missing link, and

Lucy went into most of the museums and textbooks. The models made Lucy look like a chimp from the neck up, and a human female from the neck down, but hairy like a chimp and 3 feet, 7 inches tall. Evolutionary paleontologist Donald Johanson claimed that Lucy had 40% of a complete skeleton, but most of the bones were fragments. Therefore, they did not discover 40% of an entire skeleton; no, they discovered fragments and pieces of multiple bones. If those bones had been complete, then they would have almost 40% of a skeleton. Potentially, this is stretching the truth to add credibility. One of the problems when someone is an atheist is that absolute morality from the Bible is gone, so morality becomes relative. For example, since they are convinced of evolution, then stretching evidence to prove what is true is not wrong, but becomes helpful. There is some healthy skepticism regarding how widespread the bone fragments were from different dig sites scattered around the area, even as far away as 1.6 miles, and at different strata depths spanning 200 feet (Wikipedia). They find a bone fragment 1.6 miles away and conclude that this bone fragment belongs to that other bone fragments. Then, these excavators found a partial humerus in a different gully as well. Their excuse was that since none of the bones were duplicates, then it had to be from one creature. Some evolutionists claim that Johanson found the bones much closer together and that when he said he found the knee 1.6 miles away that he was referencing a second knee. Look, even if the bones are found closer together, this is still not a problem for creationists. Why? Because some human species are bowlegged, some are knock-kneed, and some primate species have more of a bipedal lower extremity for climbing trees. This doesn't prove evolution; it still takes the imagination of the evolutionists to spin the evidence to support their biased beliefs of what happened millions of years ago, which is not observable.

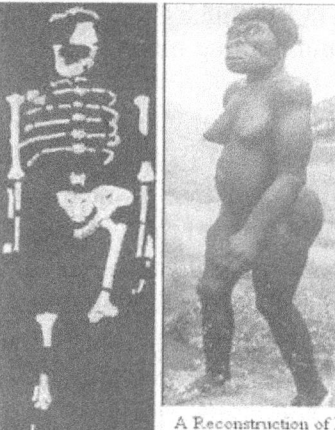

A Reconstruction of

Donald Johanson said that the overwhelming subjectivity of wanting to find the missing link created a bias that he got tunnel vision. In his words, he said (Johanson and Edey, 1981, p. 257–258.),

> There is no such thing as a total lack of bias. I have it; everybody has it. The fossil hunter in the field has it. If he is interested in hippo teeth, that is what he is going to find, and that will bias his collection because he will walk right by other fossils without noticing them. . . . that is a healthy bias . . .other bias were not so healthy . . . there is a strong urge to learn more about where the human line started. If you are working back at around three million, as I was, that is very seductive, because you begin to get an idea that that is where Homo did start. You begin straining your eyes to find Homo traits in fossils of that age . . . Logical, maybe, but also biased. I was trying to jam evidence of dates into a pattern that would support conclusions about fossils which, on closer inspection, the fossils themselves would not sustain.

I can sympathize with Johanson; we have all been there and done that on some level. However, that doesn't lessen the devastation that Lucy brought against truth. A lot of destruction against truth came about from the forefathers of evolution, and their errors in understanding still permeates academia today. "Lucy" is not a transitional fossil or a missing link.

Authors Stern and Susman, who critiqued Lucy (American Journal of Physical Arthropology, May 2005), said, "There is no evidence that any extant primate has long, curved, heavily muscled hands and feet for any purpose other than to meet the demands of full or part–time arboreal life." Meaning Lucy has long curled hands and feet for tree climbing. In addition, Lucy has a locking mechanism in the wrists for knuckle walking, which humans do not have. A human knee has an angle of 9° genu-valgus, this allows us to walk upright without waddling from side to side to shift weight. However, a chimp and gorilla, both have a 0° genu-valgus, and this is why they heavily waddle from side to side when they walk. So, the narrower the line for which a creature walks on, the greater the angle of genu-valgus to assist them walking on a narrow line. For example, to walk a tight-rope, it becomes easier with a greater knee angle, such as walking on tree limbs. And that is exactly what Lucy has, a greater angle at 15° genu-valgus angle, which is ideal for being up in the trees walking on limbs. Thus Lucy was unequivocally a knuckle walking, tree climbing, limb walking primate, and not a transitional fossil. Dr. David Menton goes into great details in "Lucy: She's No Lady!."

When Stern and Susman studied the pelvis of Lucy, they said, "The fact that the anterior portion of the iliac blade faces laterally in humans but not in chimpanzees is obvious. The marked resemblance of Lucy to the chimpanzee is equally obvious. . .It suggests to us that the mechanism of lateral pelvic balance during bipedalism was closer to that in apes than in humans."

Whenever an evolutionist wants to prove evolution, they will invariably cite Lucy as Exhibit A, yet Lucy is a spin on the evidence to support a paradigm. Lucy is just another primate, perhaps extinct, and nothing more.

Coelacanth:
This was a fish that was thought by evolutionists to be the first fish that walked out of the ocean and eventually evolved to other creatures that evolved to humans. This fish was said to be extinct 70 million years ago, until a fisherman caught a coelacanth. This is just one of several misinterpretations based on preconceived doctrine by evolutionists that happens to be wrong. Evolutionists believe that for 70 million years, the Coelacanth has not changed one stitch, yet all other creatures evolved. Their excuse is that the Coelacanth was caught in a protective non-changing bubble. http://www.cecwisc.org. www.sixdays.org.

Archaeoraptor, Fake dinosaur-bird ancestor:
Evolutionists believe that dinosaurs evolved into birds because they both have some similar aspects. Both lay eggs, have porous bones, and about 20+ things, such as clavicles, large orbits, teeth, a scapulae, and so on. But so do a lot of other animals. A missing link was needed to shore up this belief, as there is no evidence of a transitional fossil supporting this claim. Evolutionary archeologists needed to find a fossil proving this hypothesis and set out to find that elusive transitional fossil to support the claim that dinosaurs evolved into birds. By the way, dinosaurs laid, as do reptiles today, a soft leathery egg. On the contrary, birds lay calcium-rich hard-shelled egg. From 20 feet away, it seems that they lay similar eggs, but up close they do not.

Saving the day for evolutionists was the discovery of a transitional fossil called Archaeoraptor that was "found" in China in 1997. Published in *National Geographic* magazine in October 1999, this fossil was heralded as the missing link, proving dinosaurs evolved into birds.

Oh, the tangled web we weave. In the year of publication in *National Geographic*, there was a CT scan performed three months prior to publication, and the results indicated that the find could be a fraud, but decisions were made to keep that information silent. The lead team investigating this fossil, submitted a manuscript titled "A New Toothed Bird with a Dromaeosaur-like Tail" to the journal *Nature*. The journal reported back that *National Geographic* was already prepared to lead with Archaeoraptor as the cover story and there wasn't enough time for peer review. Later, it was uncovered that Archaeoraptor was a fraud, and someone had glued three to five slabs together to make one alleged transitional missing link fossil to dupe the public. After publication, *National Geographic* said that the fossil may have been a composite instead of calling it what it is, an outright fraud.

Let me help you put your mind at ease about whether dinosaurs evolved into birds; no, they didn't. Even the exposure that Archaeoraptor was a fabrication will not persuade some evolutionists who are convinced and continue to use this fraud as a transitional fossil. There was a fossil here, but it was a bird, not a transitional fossil. Don't get caught up in the fact that it has teeth; some birds have teeth, and some don't. Some reptiles have teeth, and some don't. Evolutionists proclaim that dinosaurs needed feathers, not to fly, but to keep warm. The evolutionary process of those feathers forming into wings evolved later, such as in archeopteryx. The traits of birds accumulated over time. A tree of evolution has the following possible steps: lobe-finned fish, primitive amphibian, most primitive reptile . . . a theropod dinosaur (with "**protofeathers**") . . . a maniraptoran dinosaur (velociraptor), an avian dinosaur (most primitive known bird), Archeaoraptor (e.g., the above forged fossil) . . . a modern bird (pigeon). First off, all the above names are man-made, and none of the fossils come with labels.

Protofeathers are halo-like structures adjacent to a couple of reptile fossils that evolutionists interpret as burgeoning feathers. And this accelerates evolutionists into believing that dinosaurs evolved into birds. Are there reptile fossils with "protofeathers" that represent the first stage of scaly reptiles having feathers? No. The fossils do not represent protofeathers—the pattern in the fossil is from dinosaurs that had some hair, and others are from the scales fraying from the fossilization process. During the fossilization process, the specimen undergoes tremendous pressure from the soil above, this causes the fraying patterns of the reptile's scales, and is not indicative of the beginning process of feathers. And since some reptiles today have hair in addition to scales, then the fossils may be capturing this hair. This is because hair, feathers, scales, nails, claws, horns, hooves, and so forth are made up of keratin. And it's the DNA that determines the molecular shape of the keratin to determine the end product. The evidence doesn't indicate that reptiles evolved into birds. The Bible records that birds were created on the fifth day and reptiles were created on sixth day. Therefore, theistic evolutionists are calling the Bible, the Word of God, 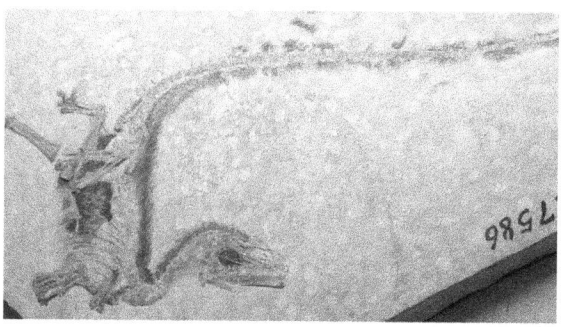 wrong and proclaiming man's (a creature created by God) hypothesis of dinosaurs evolving into birds correct. *Photo credit: Wikipedia.*

Unfortunately, the damage is done. Many people today believe dinosaurs evolved into birds and cite Archaeoraptor as proof. How many evolutionists became invigorated by this claim, and how many rejected the Bible as a result of this lie? I have talked to many theistic evolutionists and atheists who believe in evolution and cite fraudulent discoveries as proof, and they are delightful people, but that doesn't mean their beliefs are correct. The evidence stands for all to view and judge for themselves. When you see a full-scale fabrication in a museum, don't assume that it is true. Search out the evidence and see for yourselves what information or evidence was used to conclude an evolutionary link. Was it a single tooth that a whole theory was built upon? Was it bone fragments from miles apart and at different strata that led to a conclusion? Search for yourself, use your own mind, and don't rely on what others tell you. There is ample evidence that mankind was once larger before the flood.

Review: Evolutionists can proclaim their belief, and they can shout louder than others, but at the end of the day, their missing link fossils are either forgeries, misinterpretations, or misleading claims. From the fraud of Haeckel's embryo drawings, to Neanderthal man, and so forth, their claims of finding missing links that prove evolution are wrong and only reveal biased beliefs that govern their objectivity. Claims come and go, and theories will be modified, yet the Bible remains unchanged and stands the test of time with the phrase that each kind (~Family) will produce

"according to its kind." Both sides of the debate, creation and evolution, have admirable followers, and there have been forgeries, lies, and misinterpretations through the centuries, yet the Bible remains the same. You see, the Bible is not defined by followers. When a few wolves in sheep's clothing commit evil and hide behind the banner of Christianity, they are on their own and not following the Bible, which still stands as truth. But when an evolutionary forefather commits a forgery to build their doctrine upon, then that is equivalent to Jesus committing evil to promote the Word of God. When Darwin builds a theory about spontaneous life from nonliving material because he has wrong information about the complexities of the cell, then that would be equal to Jesus teaching error because He was lacking knowledge about biology. What the forefathers of evolution did to promote their hypothesis matters, just like it matters what Jesus did to authenticate the Bible.

Group Discussion:

1. In regard to many of those "transitional" fossils you may have heard about in school, how did they affect your faith?

2. As your knowledge increases in truth, how do you see this affecting your faith? Read Hosea 4:6a and Romans 10:17 for application.

Chapter 23
Light

The speed of light is 186,000 miles per second. This is incredibly fast, but considering the vast size of the universe, it still takes light a long time to traverse the galaxies and universe. The sun's light, though it is far from Earth at ~93 million miles, takes 8.3 minutes to reach us. Now, astrophysicists hypothesize that though the sun's light only takes ~8 minutes to reach the earth from the sun's surface, those light particles took millions of years to reach the sun's surface from within the core. This is because the sun's density is immense. Evolutionists argue that this is one reason that young earth creationists are wrong, because when we see the sun's light, it's not just 8 minutes old, it's actually millions of years plus 8 minutes old because that is how long it takes the light to travel from the sun's core to the surface to earth.

How does a creationist handle this? For one thing, the hypothesis that it takes light millions of years to traverse the sun's core is a theory; it's not a proven fact. And even if this hypothesis is true, God didn't create Adam and Eve as babies; He created them fully mature. Likewise, God created the sun fully developed, which means that immediately after the matter that would eventually form into the sun was created on the first day, it was spinning and coalescing matter together until the sun glowed hot on the first day of creation. By the fourth day of creation, the sun coalesced enough matter that it ignited. The light emanating from the surface would be traveling to earth and not having to wait millions of years for light to leave the core. Therefore, since this occurred 6,000 to 10,000 years ago, we are currently viewing light that started just below the crust and worked it's way to the surface and then to us from creation; we do not view light from the sun that is millions of years old and worked its way from the core. Adam and Eve viewed the same sunlight; however, it was 6,000 years younger, and they viewed the light through the filter of the canopy.

The Bible records that waters, specifically the "deep," which suggests salt water, was abundant in the small universe on the first day. And 2 Peter 3:5 clearly states that the earth was made from water and through water, but Holman Christian Standard Bible and the Aramaic Bible in Plain English both interpret that both the heavens and the earth were made from the water and through the water. Thus, by means of the sun's spinning core of liquid metals, this would cause the water, H_2O, to separate by the process of electrolysis into H_2 and O_2, at a 2:1 quantity ratio. And given the immense pressure and heat, the diatomes of H_2 and O_2 converted to plasma. Therefore, I contend that the sun consists of an outer lighter layer of H_2 plasma and an inner heavier layer of O_2 plasma. The implication is that with the layer of O_2 plasma, then fusion doesn't need to only occur at the core and take millions of years to reach the surface, fusion may occur more distally and, thus, take less time to reach the earth.

Well, what about other stars and other galaxies that are estimated to be billions of light-years away? Well, let's remember that a light-year is a measurement of distance and not time. A light year is the distance light travels in one year. By using triangulation calculations, we are only accurate up to 400 light-years for star distances; beyond that, we switch to utilizing starlight brightness to estimate the distance. This results in a compounding of the error rate to determine distances because we use nearby stars to determine the distance of stars farther away, which means the error rate increases the farther away the star is that is being measured. Therefore, there are potentially large errors in the number of light-years for the farthest distal galaxies. Notwithstanding, I think our universe is even larger than we can fathom. Don't get too caught up on reading that "x" galaxy is "n" distance away from the earth and the notion that this proves the universe is 14.6 billion years old. Why is this not a problem?

For one, humans are able to accelerate the speed of light impulses to 300 times the current velocity to 55.8 million miles per second. Dr. Lijun Wang of the NEC Research Institute of Princeton has successfully increased the speed of light by 300 times. Just give mankind more time, and they will discover that they can speed up the velocity of light even greater.

To add to this theory, some scientists suggest that light has a decay element to its velocity. For example, in terms of the electromagnetism of planetary bodies, they have an ever-decreasing electromagnetic field. Dr. Troitskii wrote that "the speed of light could have been 10 million times faster in the past compared to what it is today" (Astrophysics and Space Science, 139, 1987 p. 389–411). This seems plausible, since everything in the universe is winding down and tending toward decay. Light could also tend toward decay.

Review: Light immediately started traversing from a star's surface instantly upon creation. Light didn't have to travel from the core for a million years before the star began providing light. The speed of light can be accelerated 300 times the current speed. If the speed of light has a decay process like everything else in the universe, this too would solve the problem with the distances.

Measuring the distances of stars today doesn't mean they have always been at that distance. There is a force in the seemingly empty spaces of space that is forcing the universe and each galaxy to expand away from each other at an ever-accelerated rate. This means that the universe is expanding faster and faster and not at a constant speed, indicating that the universe was significantly smaller in the past. This suggests that light didn't have to travel as far to reach the earth from its source because it started off closer in a much smaller universe. The Bible has several verses that discuss the universe being stretched out at creation and one that suggests that God continues to stretch out the universe. God "stretch**ed**" out the universe at creation on the second day of creation (Gen. 1:6–14-expanse, Isaiah 51:13). Also, according to Isaiah, God "stretch**es**" out the universe (40:22).

From the Biblical narrative, it appears that our solar system, though located at the edge of the Milky Way, could be near the center of the universe, as the Genesis record is from the perspective of Earth. This is just to illustrate that the earth is not at the edge of the universe, but is closer to the middle for distance purposes. In fact, cosmologists observe that there are equal numbers of galaxies in every direction in the universe. The best way to shorten the distance for light to travel from one edge of the universe to earth is to place Earth in the middle of the universe; now the distance is cut in half. But we still have billions of light-years to work with as a creationists. How does a Big Bang theorist (basically all evolutionists) contend that the universe is larger than the allotted time from the Big Bang today? Even atheists have an insufficient amount of time to explain light traveling from the farthest distant galaxy to reach the earth. Their solution is that the universe expanded faster than the speed of light at the beginning, and then the expansion leveled off. This accelerated expansion is called universal inflation (by Alan Guth, Theoretical Physicist), where the universe expanded by a "factor of 10^{90} in one millionth of a billionth of a billionth of a second," according to Lawrence Krauss, a theoretical physicist and cosmologist. This is like a grain of sand swelling larger than our sun and doing so faster than the speed of light. Creationists are interested in the notion of universal inflation, but we disagree when this rapid inflation occurred and the duration of the inflation. To a creationist, universal inflation, as far as expansion rates during "Let there be an expanse," fits well with creation and is simply a calling card that God expanded the universe on the second day of creation. Evolutionary cosmologists believe that so much happened during the first second of the beginning of the universe, the Big Bang, that Dr. Plank categorized the brevity of time in Plank units. Creationists view the universal inflation to have occurred at the beginning of the second day of creation, and lasted many hours. Utilizing an evolutionary construct of universal inflation and the Big Bang, creationists are successfully able to argue a young universe. There is a bit of irony there. What man has intended to explain away God by natural processes winds up furthering support of a young universe. This seems a fulfillment of Romans 8:28, "And we know that God causes all things to work together for good to those who love God."

Since both sides accept the accelerated expansion hypothesis, then where do they differ? It is in the amount of matter and volume of the universe "In the beginning" before the accelerated expansion, and the duration of the universal inflation. Evolutionists contend that there was zero matter because all the matter existed as energy and was compressed into a dot. Creationists contend that after God created the matter on the first day that it was compressed enough to allow water to exist in four forms (ice, liquid, gas, and plasma). And God began expanding the universe during the second day of the Genesis creation account. This universal inflation concept is a plausible explanation of how we view stars from the earth in what appears to be a shorter period of time than what the average speed of light allows in a young universe hypothesis.

Creationists say that God created everything at a mature age free of defect, such as an adult Adam and Eve and tall trees, but this does not mean that God created anything with an alleged history. God does not deceive, and God cannot lie (Titus 1:2). For example, God didn't create tree rings or belly buttons for Adam and Eve. Nor did He fake a supernova explosion, so humans could perceive the remnant light. The supernovas simply exploded when the universe was much smaller. Therefore, their light didn't have to travel as far to reach Earth. And there is another solution in physics that also adds to what we have talked about so far, and that is space travel. Physicists have theoretically shown that mankind may be able to travel 10 times the speed of light in a type of warp drive. The theory is based on the fact that space expands and contracts; if the space in front of a ship is contracted and the space behind the ship is expanded, then a warp bubble is created, so it has been theoretically shown that we can move up to 10 times the speed of light. Well, if man may be able to do this with a ship (and mankind accepts that the universe expanded faster than the speed of light in the beginning), and if an expanding universe traveling faster than the speed of light also has the same warp bubble (with contracted space in front of light and expanded space behind the light because of a violent expansion), then light could have traveled at least 10 times the speed of the universe. If you are overwhelmed at this point, just know that scientists accept the notion that objects can travel 10 times faster than the speed of light. If light itself was that object, then it too could have traveled faster than 186,000 miles per second. If you are still confused, think of a surfer riding a wave; this explains the concept of light traveling along the wave of universal expansion.

Review: The earth may very well be in the galaxy that is in the center of the universe. This potentially cuts the distance light has to travel to reach Earth by half. Some scientists accept the notion that the universe expanded faster than the speed of light in the beginning. Since scientists believe that the universe already did this, then it certainly should be acceptable for an all-powerful God to do the same. Since it has theoretically been shown that a spaceship can travel 10 times faster than the speed of light, then what would limit God to do the same with light? Since man can accelerate the speed of light, then God can accelerate the speed of light.

There is another aspect to the speed of light that is important. We have all been told that the speed of light is a constant at 186,000 miles per second (3 million meters/sec.) in a vacuum, but it turns out that the speed of light can be accelerated 300 times its speed, as discussed above, and now physicists are able to slow down the speed of light to less than the speed of a bicycle rider to a couple of miles per hour. Physicists are able to direct a laser beam of light into a medium called a Bose-Einstein condensate, and this medium is dense enough to slow the speed of light down to a couple miles per hour without any loss of information (Lene Vestergaard Hau, Harvard University). This is a reduction in the speed of light by a factor of 20 million.

Therefore, if light can be slowed down by a factor of 20 million and sped up by a factor of 300, then why couldn't an all-powerful God perform the same acceleration of light? The creator of light may manipulate light how He seems fit. It's ironic that evolutionists use this light argument against young earth creationists while touting the wonders of humanity at the same time for discovering that the speed of light can be slowed down or sped up, but then they limit the ability of God to do the same.

They also use an argument against young universe creationists that suggests that there is not enough time for light to reach the earth from seemingly so far away. Yet, even 14 billion years is not enough time under their own parameters that they have established to explain a universe that is 93 billion light-years across and growing with each invention of a more powerful telescope. Think about that. There are some plausible explanations of how light has traveled so far in such a short amount of time for young universe creationists and for evolutionists. In the seemingly empty spaces between galaxies, vacuums in space are rarely found. Even the so-called empty spaces between planets and galaxies are filled with dark energy. And this dark energy pushes galaxies apart from each other at an ever-accelerating rate. This means that the distances measured today do not measure the distances that light had to travel in the past.

The Bible implicitly and at times explicitly records that God created matter in a small universe, with trillions of areas that immediately began coalescing (spinning and gathering matter together), until enough matter gathered together for magma to glow (light) at the middle of the first day. And as a result, pressure, heat, and potential energy built up until the Big Bangs, when God began stretching out the universe on the second day of creation and then enough coalescing matter ignited to form the sun and stars during the fourth day of creation. Immediately, light from the middle of the first day, from burgeoning stars, started traversing the distances of the small universal space. This scenario results in light traveling from the stars' initial location, which was much closer to the earth and not nearly as far as light has to travel today. Therefore, the light, from the glow of the burgeoning and coalescing stars on the first day of creation, had no problem traversing the entire universe because the universe was small. And the light from stars that coalesced enough matter to ignite on the fourth day of creation, had no problem traversing the entire universe because the universe had just began expanding (Big Bangs) on the second day of creation. The distances today between the earth and all other galaxies is much greater than at creation. Really, this is a simple concept. The source of light was closer to the earth at creation than today, and the distance grows ever greater with the increase in dark energy that is pushing galaxies away from each other. In fact, Lawrence Krauss said, "In 5 billion years, the expansion of the universe will have progressed to the point where all other galaxies will have receded beyond detection. Indeed, they will be receding faster than the speed of light."

The Bible declares that God finished forming the sun, moon, and stars on the fourth day and said, "Let them be for signs and for seasons and for days and years . . . and it was so" (Gen. 1:14–19). This indicates that God declared the stars were to be used as measurements of time immediately. If the stars were formed in the size of the universe today, then it would take a stars' light billions of years to reach the earth, and God would have to wait for His Word to be fulfilled. But God supernaturally used natural means to cause light to traverse the small distances of space before creation was finished.

Dr. Raymond Chiao, a professor of physics (University of California at Berkeley), has reviewed Dr. Wang's work on accelerating light by 300 times and conducted his own experiments that indicate simultaneous multiple localities. Photons, the particles that constitute light, seem to apparently jump between two points separated by a barrier in what appears to be zero time. This is called "tunneling." Should we be shocked by this? Was it not God who caused Philip to appear in one city sharing the gospel and then, poof, he was snatched away and appeared in another city (Acts 8:39–40)? Was it not God who snatched away Elijah (1Kings 18:12) and Ezekiel (Ez 3:12,14, 8:3) and Paul (2 Cor. 12:1–6) to heaven and then back potentially without any time passing? Was it not God that walked through walls at His resurrection and just miraculously appears in front of Joshua (Joshua 5).

Why does mankind limit God's abilities? Yet, as mankind advances in knowledge, people are able to perform things that they use to mock God about. People mock believers who suggest that their God accelerated the speed of light for us to see distant stars or slowed the light in Joshua's day for a battle, and now, humans can accelerate and decelerate light. People mock believers that accept a young universe because of the distance of stars, yet they accept the notion that the universe expanded faster than the speed of light near the time of the Big Bang. People mock believers who accept that God created the sun, moon, and stars, yet some people believe that all the stars and planets and the universe evolved from nothing—that nothing exploded from nothing and formed everything. In this "nothing" at the beginning, at the Big Bang, there was allegedly no matter, but only massive amounts of pure energy—and when the Big Bang exploded, that energy eventually formed into a more balanced state of matter and energy and then coalesced into the galaxies as we know it.

Evolutionary scientists try to explain away God via natural processes, and it seems at times as an attempt to make it impossible to believe in God, but science brings us back to believing in God. In a sense, the more we learn about science, the easier it will be for people to believe in God again. The old adage was that a little bit of knowledge can be harmful; well, people have perished for a lack of knowledge (Hosea 4:6). As knowledge increases, it is becoming easier to see God. For example, some people mocked believers for accepting Jesus' virgin birth, yet mankind has achieved the same ability with artificial fertilization. Many mocked God for creating mankind in His image yet have accepted cloning. Many mocked God for the talking snake and donkey, yet they will buy a parrot to talk to. Many mock the resurrection of the dead, yet they learn CPR and pray for the doctors to keep trying to revive their loved one. The double standard continues. Atheistic evolutionists will mock all the way up until the heavens open up and meteors fall on them (Revelation 6:15–17) and in their time of trouble, they will say to God, "arise and save us." But God will say, "Where are your gods which you made for yourself? Let them arise, if they can save you" (Jeremiah 2:27–28).

Both evolutionists and creationists agree that the universe was smaller in the beginning. Using an evolutionary concept that the universe was able to expand faster than the speed of light in the beginning, then a much smaller universe allows light to reach earth from the stars in a short amount of time, and as the universe expands, now the light can travel in both directions to keep the stream of light continuous from the source to the earth. Therefore, their light didn't have to travel as far to reach Earth.

Evolutionary astronomers believe and teach as fact, that it takes billions of years for stars to evolve from a red giant to a white dwarf. This is by faith alone, not science. In fact, ancient astronomers documented that Sirius—now a white dwarf—was redder than Mars. Paul Ackerman, author of "It's a Young World," listed the ancient astronomers that recorded Sirius as a red star: Egyptian hieroglyphs (2000 BC), Cicero (50 BC), Seneca, and Ptolemy (150 AD).

Review: Since mankind can alter the speed of light, then God—who created those clever humans and created the light in the first place—can accelerate light. Mankind has discovered that the universe is expanding, which means that when God created the stars in the galaxies, they were closer to the earth, and their light didn't have to travel as far to reach us. Scientists accept the theory of universal inflation, which expanded the universe faster than the speed of light, since scientists speculate about the extreme brevity of this duration (Planck time) for the inflation, then if the duration of the universal inflation lasted a few hours longer, this would result in a young universe.

Group Discussion:

1. Have you ever doubted the Word of God because you were told that the distant starlight couldn't reach earth in a literal Genesis creation?

2. Were you ever mocked for believing in God's Word? Read 2 Timothy 3:12 and self-evaluate whether your life is a threat enough to Satan for him to consider persecuting you.

3. How has this chapter affected your faith in the Word?

Chapter 24
Humans Lived 900+ Years and Adaptation from Origins

The Bible records that people from Adam to Noah lived 900+ hundred years of age (Gen. 5:1–32). Then, something dramatic happened during the life of Noah, and from then on, the length of life started declining. Let's look at the life span of the first humans on earth, with the name and number of years lived (in chronological and genealogical order): Adam 930 (Ca. ~4004 BC, roughly ~6,000 years ago), Seth 912, Enos 905, Cainan 910, Mahalaleel 895, Jared 962, Enoch 365 (God took him early to heaven; he never died), Methuselah 969, Lamech 777, Noah 950, (at age 600 Gen. 7:11—**the catastrophic global flood,** ca. 2348 BC**).** Shem **600**, Arphaxad **438** (Gen. 11:10–32), Salah **433**, Eber **464**, Peleg **239** (In his days the earth was divided—the Tower of Babel), Reu **239** (around the time Job lived), Serug **230**, Nahor **148**, Terah **205**, Abraham **175** (Gen. 25:7)—Noah dies 350 years after the Flood and sees Abraham, Isaac **180** (Gen. 25:28), Jacob **147** (Gen. 47:28), Joseph **110** (Gen. 50:26).

There are several important points about the list of ages of the patriarchs. One is the enormous time that each person lived before the Flood. Second is the clear reduction of ages immediately after the Flood, but notice that the ages gradually decline as mankind adapts to the changes. Please note, that when humans once lived 900+ years, though their vertical height stopped growing, human facial bones never stop developing. The jaw bone and the eyebrow ridges continue to protrude as one ages. This may account for some fossils that evolutionists contend are missing links. This facial bone continued development would be present for each creature that lived 10 times their current lifespan, as humans once did.

People lived 900+ years of age? Yes. For one thing, life was not cut short via disease or sickness. When Adam and Eve sinned (Gen. 3:23), the affects of their sin would be immediate (spiritual death) in one sense and delayed (physical death) in another sense. Sometimes the cost of our sin is immediate, and sometimes the cost of our sin is not seen until later or even until our children bear the inheritance of our sins. Adam and Eve paid an immediate cost for sin with spiritual death and the loss of the Garden of Eden, pain in childbirth, hard work for food, and so on. But some effects of their sin would take generations to see, such as one son killing another son, judgment with a global flood, and their distant heir (Jesus, Luke 3:23–38) having to die on the cross to pay the penalty for all sin.

There was no mauling from animals to shorten one's life. Yes, predators lived at that time, but they had not adapted as a result of sin and disobeyed God's command to only eat vegetation. Adam and Eve, and all animals were herbivores (Gen. 1:30–31). There was no food poisoning, no harmful bacteria in the food they ate, no mold, and no fungus in the food to make one sick and die because they simply plucked ripe fruit and nuts off the trees in the Garden of Eden. Any fruit that had too much bacteria would fall to the ground and be accompanied with the normal discoloration and odor. Yes, bacteria, viruses, and fungi existed, but potentially harmful strains had not yet adapted to external stimuli, such as high-energy rays from space, and internal stimuli, such as sin. There was no obesity with early humans. How? There was no fast food, no soda, no fried food, no processed food, no canned food, no candy, no ice cream, and so on. People just ate natural fruit, vegetables, nuts, and grains. This meant that life was not cut short from maladies associated with diet.

In addition, the oxygen concentration was 50% higher than today. This caused life to thrive. Also, the effects of gravity were reduced before the Flood. It's not so much that the mass of the earth has increased that much over time; it's that there was a canopy of salt water surrounding the atmosphere and creating a buoyancy effect. This buoyant force reduced the net effect of gravity by about ~20%. Therefore, there would be less strain on the cardiovascular system to circulate blood throughout the body, which prolongs life.

God mostly uses natural means to carry out His judgment, for example, in the case of the Flood. As a result of the Flood, mankind's life expectancy gradually decreased from 900+ to around 90 years.

How? The flood significantly covered the land that once produced oxygen, so this physically reduced oxygen production and reduced the oxygen concentration over time. Life is dependent upon oxygen, so with reduced oxygen, life doesn't thrive as it once did. The Flood removed the canopy of salt water, which removed the buoyancy effect. Therefore, this increased gravity puts extra strain on living beings. A greater strain on the life support systems of living creatures results in a greater reduction in life expectancy. Also, the loss of the canopy resulted in more harmful rays of the sun affecting life on the planet. This resulted in the creation of deserts and large global changes in temperature and polar ice caps. This led to tornadoes, hurricanes, and other storms that shortens life as well.

Can mankind live 900+ years again? God's judgment resulted in the fact that mankind would no longer live to extreme ages. But, this is not how it will always be. For God has decreed that He will undo the curse of sin and restore the longevity of humans again. When? It will be immediately after the battle of Armageddon, when Christ returns and crushes evil. Then, humans will live long lives again. Isaiah 65:20: "No longer will there be in it an infant who lives but a few days, Or an old man who does not live out his days; For the youth will die at the age of one hundred and the one who does not reach the age of one hundred will be thought accursed." This is describing a future time when living beings will live so long on average that if a few do happen to die at the age of 100, they will be considered dying as a youth. This is not just humans, but all life forms will live a long life.

The event that single-handily changed the longevity of humans was the global flood of Gen. 7. How? Let's go ~1,650 years before the Flood and look at the creation account of Gen. 1:6–9: "Then God said, 'Let there be an expanse.'" *Expanse* means to expand like a blacksmith would hammer out metal, with a violent force. This parallels the violent expansion concept of the Big Bang. The word *firmament* is used in the NKJ, and it means the same thing. Simply put, it's an opening, or an expanse formed by violent force. "Let there be an expanse in the midst of the waters, and let it separate the waters from the waters." This means that God made an opening in the midst/middle of the waters. *Waters* is plural to represent either abundance or the four forms of water: liquid, solid, gas, and plasma. Verse 7: "God made the expanse, and separated the waters which were below the expanse from the waters which were above the expanse; and it was so." Notice that the *expanse* is singular. Verse 8: "God called the expanse heaven." Verse 9: "Then God said, 'Let the waters below the heavens be gathered into one place and let the dry land appear'; and it was so, God called the dry land earth, and the gathering of the waters He called seas." Therefore, we know that this expanse/heaven is the atmosphere. Also, notice that God now uses the plural form of *heavens*. Therefore, God, utilizing violent force, stretched out another expanse/heaven on the second day of creation as well. And the expanse of the universe (the Big Bangs), is what caused our atmosphere to form. Although the sun, moon, and stars are not fully formed (ignited) until the fourth day, God initiated the expansion of the universe that caused the expanse of the atmosphere. There is no reason to think that God wouldn't create the universe in the same manner as He created the atmosphere by separating the waters above from the waters below. A tightly packed universe, with a shell of thick ice surrounding it, would increase internal pressure as coalescing stars vaporized water. As each coalescing future star rotated liquid metal, this caused electrons to flow and build up a charge. This unstable build up of electricity and electromagnetism would have caused a violent lightning discharge to spread throughout the densely packed universe, thus, instantly vaporizing water and increasing the pressure. When the internal pressure exceeded the tensile strength of the exterior shell, then the universe violently expanded (trillions of Big Bangs); this sudden release of pressure in the expanding universe caused the water surrounding the earth (includes every planet and moon that had water surrounding it) to instantly vaporize into gas. This process is called cavitation, and it is seen every time a soda bottle is initially opened and pressure is released. Potentially, every planet in the universe may have had waters above and waters below its atmosphere at its origin. However, conditions were not favorable for retention of their atmosphere or of their waters.

Review: All life once lived ~10 times their current longevity. God judged the wickedness of all life, which resulted in a severe reduction in the number of years lived. God will restore this longevity. What is the purpose of studying Gen. 1:6–9? The key is that God made the atmosphere in between the waters. The waters below became the seas, and the waters above became the canopy of salt water surrounding the atmosphere.

What is the purpose of this canopy of water? It shields the earth from the harmful effects of the sun to prevent deserts and to allow lush vegetation to grow there instead. It creates uniform global temperatures to prevent polar ice caps and allows lush vegetation to grow there instead. Since the canopy rained upon the earth to form oceans, then prior to the Flood, only smaller seas existed. And this allowed for lush vegetation to grow instead. More vegetation equates to more oxygen. More oxygen equates to longer life span.

How do we know this? Let's look at this from a negative perspective and see how life suffers with a lack of oxygen. There are many terms associated with a lack of oxygen: pale, fatigue, avascular necrosis, ischemic necrosis, myocardio-infarction, embolism, death, numbness, vertigo, and so on. Oxygen is vital to life. Without oxygen, we die within seven minutes, and within three minutes without oxygen, we could have brain damage.

Now let's look at this from a positive perspective and confirm that life thrives in an abundantly rich oxygen environment. Arizona University performed a series of tests. They raised insects in hyperoxia (high O2) and hypoxia (low O2) to determine if there was a change of life for insects. They found that all the insects in their study were positively affected when raised in an hyperoxia environment, and all suffered some ill effect when raised in an hypoxia environment. The positive effects on the insects living in hyperoxia were that most grew larger and grew faster than normal. The ones that didn't increase in size still had smaller tracheal tubes, which allowed for larger muscles, tendons, ligaments, internal organs, and so on. Therefore, they were super insects that were faster and stronger and had greater endurance.

Review: We have evidence that the canopy of water resulted in greater vegetation quantity, and that vegetation produced greater oxygen concentration in the atmosphere, and greater oxygen caused life to thrive. Therefore, the canopy of water caused life to thrive.

DNA, RNA, genes, and cells are adversely altered from ultraviolet rays, cosmic rays, X-rays, and gamma rays. These high-frequency light particles and wavelengths have an adverse effect on our lives and affect the accuracy of genetic duplication of DNA. Though genes have protein markers that correct errors and delete unwanted errors, there are occasional random, unguided errors in the gene code. These random, unguided mutations always adversely affect functions, the host, and efficiency, and they never enhance functions or the host, and they never initiate a new function. It turns out that water is a barrier to limit those harmful high-energy rays. Well, the canopy shielded against those higher energy light waves from adversely affecting DNA, RNA, genes, and life on earth. Before the Flood, the offspring of each generation from Adam and Eve were more accurate genetic copies of the first humans on Earth. After the Flood and the loss of the canopy, there were increased errors in genetic duplication with each generation. This mutated gene reduces the likelihood and sometimes prevents the host from having offspring, as all creatures are preprogrammed in their DNA to select the strong and not the weak and to select the genetically superior and not the genetically mutated. In addition, those with mutations are usually weak, so predators and climate kill them. Also, the genetically mutated are sometimes sterile to prevent passing on their mutated gene. There is a huge and impossible uphill climb to support evolution, as evolution not only needs one example to show how mutated DNA enhanced a host, but it needs trillions upon trillions of beneficial mutations to occur for life to evolve into the complexities we see today. Yet, ask any brilliant evolutionist to show one beneficial mutation, and they will pause and say, "You are asking the wrong question, since the observable changes, such as fish to amphibians or primates to a transitional phase of humanity, occurred hundreds of millions of years ago." This is an attempt to redirect you away from the lack of evidence of beneficial random mutations; it's smoke and mirrors and requires a leap of faith.

NASA has figured out that a canopy of water of a sort can protect astronauts from harmful high-energy radiation and block those harmful rays from penetrating the hulls of spaceships. NASA is proposing to surround the hulls of spaceships with water to protect humans and vital instruments on long durations in space.

Review: The canopy of water indirectly caused more accurate copies of genetic information by limiting high-energy rays from space that can cause harmful mutations of DNA information.

Not only does water limit harmful high-energy radiation from penetrating the hulls of spaceships and onto the earth's surface, there is another benefit of the canopy that is at the crux of the matter for longevity of life. But to get there, we need to lay a foundation of DNA. The DNA of all life is double helical and is composed of long threads of digital information of As, Cs, Ts, and Gs. DNA allows for the formation of chromosomes, which are located inside the nucleus of a cell. At the tips of the chromosomes are caps of DNA information that protect the rest of the chromosomal strand from fraying or splitting, sticking to each other, or losing information, all of which are harmful. These caps of DNA also allow cells to divide. These caps of DNA are called telomeres. For this reason, some teachers of genetics have likened them to the plastic tips on short shoelaces.

Unfortunately, there is a catch to these telomeres. Each time the cell divides, the end caps get shorter and shorter until they are no longer able to be reduced for cell division, and cell division stops. This causes the cell to become inactive, or it dies. And for this reason, geneticists take the plastic tips of shoelace example to another level and say they are also like fuses for a bomb. There is no explosion, but old age or cancer and death are associated with the ends of the burning fuses.

What is so important about cellular division from these telomeres? Cell division is responsible for the new growth of skin, blood, WBCs, nails, hair, bones, neurons, and so forth. It also augments wound healing, pigment repair from damaged skin (from harmful sun exposure), elasticity of skin for sagging, the repair of damaged collagen (for wrinkles), and so on. Just about everything we associate with aging is directly affected by the ability of cellular division from these telomeres. Another critical ability of telomeres is that they reduce in size when the cell divides, and this prevents chromosomes from reducing in size, preventing the loss of DNA information and mutations in offspring. To clarify for evolutionists, mutations are never positive, so when DNA information is lost and a child is conceived with parental genetic information in the chromosomes that is lost, this is not proof of evolution; this is a birth defect. To counteract the shortening of telomeres, the body produces a protein enzyme called telomerase that adds bases (As,Cs,Ts, and Gs) to the ends of telomeres to extend their life span. Unfortunately, as cellular division continues, the quantity of enzyme telomerase may not be sufficient to keep up with the demands of adding bases to extend the life of telomeres. This results in a loss of ability for the telomere at the tip of a chromosome to be reduced further with cell division, and subsequently, the cell becomes inactive or dies. This results in aging of the host of the cell. Geneticists have determined a link with those that die prematurely with having shortened telomeres as compared to those living longer lives. Also, they have noticed that the production of telomerase enzymes is associated with prolonged telomeres. Therefore, having a sufficient production of telomerase enzymes could be a causal effect of the longevity of life. There is a side effect though, and that is cancerous cells. High telomerase enzyme counts have also been associated with cancerous cells. Why bring this up here in this section of the book? Water can limit radioactive energy from penetrating into the human body and causing cancer. Therefore, the canopy blocked solar radiation from space from adversely affecting life on earth and mutating DNA, mutating genes, and causing cancer and cell death. Therefore, there is a link between the canopy of water preventing harmful high-energy rays from penetrating the earth's atmosphere and altering cells, which would have altered the production of telomerase enzymes. Additionally, with the canopy protecting life on earth, there would be less damage to cells from high-energy rays, and therefore a reduced requirement for telomerase enzymes to keep the body young and repaired for maximum longevity of life.

Geneticist Richard Cawthon says that "people could live 1,000 years." If we could make a synthetic telomerase and implant the enzyme into the body, it would heal the body of almost all symptoms and conditions and delay the progression of conditions that reduce the quality of life. There wouldn't be a limit to what we could cure. It might be possible to cure cancer by using telomerase inhibitors that allow cancerous cells to continue with cellular division and grow. By blocking telomerase production to an isolated cancerous area, then the cancer cells wouldn't be able to sustain cell division and would die. Well, with a canopy of water, there would be no harmful cosmic rays, gamma rays, x-rays, and ultraviolet rays to turn normal cells into cancerous cells in the first place. This means there wouldn't be any wasted telomeres repairing destroyed collagen from ultraviolet rays, and there wouldn't be any high-energy rays that damage any tissue (which would use up the DNA telomeres caps), and therefore, there would be more protein enzyme telomerase left over for prolonged life.

As a result of cosmic energy (or any other harmful stimuli) that has been adversely effecting (mutations) the DNA for generations, then as we travel backwards in time, each generation would have less mutations. And this is what geneticist, Dr. John Sanford, has discovered. In his book, Genetic Entropy and the Mystery of the Genome, he elucidates that the DNA is entropic with each generation, which means that the DNA is becoming more disorderly or prone to more harmful errors. This is opposite of the notion of evolution, that the DNA is becoming more orderly, more complex over time.

Review: There is an association with humans that live longer and having longer telomeres DNA caps at the ends of their chromosomes. A link with more production of telomerase enzymes is also associated with the longevity of life. A source that depletes both telomeres and telomerase is high-energy radioactive rays from space. Water can block high-energy radioactive rays. Therefore, both telomeres and telomerase enzyme would positively be affected by a canopy of water that protects life from harmful high-energy rays, thus prolonging life. Due to Genetic entropy, as we travel backwards in time, the DNA in life is less mutated until we reach a pair of humans with zero mutations, we call them Adam and Eve.

There is still another benefit from the canopy of salt water. As mentioned previously, it would cause a higher atmospheric pressure on earth, which creates a buoyancy effect and reduces the effect of gravity. Changes in gravity can have profound effects on the human body. Astronauts are susceptible to bone density loss with prolonged space travel. One experiment with an astronaut on a space station revealed that he lost 14% bone mass with six months in space.

As a doctor, I have noticed that heavyset patients rarely suffer from bone density problems. In fact, when viewing radiographs, it is clear that heavyset patients have thicker bones than skinny patients. Sure, they tend to suffer from joint problems due to their extra weight, but their bones compensate for the extra weight by adding more mineral deposits. The negative of this is true as well. Older patients that are too skinny have an increased risk of thinning of the bones. The propensity toward osteoporosis increases with weight loss and increased age. So what is osteoporosis? And what do reductions in bone mass, bone density, and increased porousness mean? They are all saying the same thing. They are caused when the cortex of the bone (outside portion) and the trabeculae of the bone (inside portion) get thin.

How does this play out for our discussion? Well, before the global flood of Gen. 7, the canopy applied weight and subsequent pressure on the inhabitants of the earth. This greater weight increased the total atmospheric pressure on life on earth. This created a buoyant force that reduced the net effect of gravity by approximately 13%–25%.

How does increased atmospheric pressure reduce gravity? Gravity is pulling objects directly down to the core in only one direction, whereas increased atmospheric pressure pushes on objects on the earth from all directions, such as objects in water. When an individual is in water, the pressure from the water pushes from all directions on the body, even in an upward direction, whereas gravity is only pulling down in one direction. Therefore, there are many vectors of force from the increased atmosphere countering gravity. Thus, gravity appears weaker. But from a mathematical view, the weight of the fluid displaced by a body is the amount subtracted from the force of gravity and their mass. If the body weighs 200 pounds, and the volume of the air (pre-Flood) displaced weighs 40 pounds, then the buoyant force causes the person to weigh 160 pounds.

Is there evidence that creatures had less force on their skeletal structures before the Flood? Yes. Dinosaur bones are porous. In fact, scientists agree that the skeletal systems of dinosaurs, if they were alive today, could not support their massive tonnage because of today's gravitational environment. Why? Dinosaur bones are porous and would break under the enormous weight. Unfortunately, evolutionary scientists believe that the porous bone structure of the dinosaurs is proof that they evolved into birds because both have similar porous bone structures. I haven't heard any scientist suggest the seemingly obvious option that dinosaurs' skeletal systems were thinner in correlation to weaker gravity and therefore experienced less stress on their skeletal systems. Less strain, less weight, and less stress on the skeletal system equates to thinner and more porous bones.

Is there evidence of a correlation between the load on a creature's skeletal system and the quality/density of its skeletal system? Yes. With increased load on the skeletal system, the brain responds by adding more bone mass to compensate. This load can be from increased weight or increased activity, such as from sports. The opposite is true as well. With less weight/load and fewer activities, the brain responds by allowing a decrease in bone mass. This reduction of bone mass would be appropriate for the reduced stresses placed on the skeletal system.

Let's take the reduction of loads on the skeletal system to an extreme location, such as in the weightlessness environment of space. We have noted that astronauts lose 14% of their bone strength from just six months in space. That is a lot of bone mass. Most of the time, the bone loss is mostly reversible. Some astronauts have been reported to lose up to 30% of bone strength from prolonged space travel.

I made a living viewing radiographs (X-ray film) of patients' spines and diagnosing focal points of sclerotic white spots in the joints, where one side of a joint appears larger than the other side and represents an increase in mineral deposits. This particular increase in bone formation is usually associated with joints that don't share loads and ranges of motion. Therefore, one side of the joint has more load than the other side due to poor alignment, and the result is that the brain directs the body to build up strength in that region. The increased deposit of minerals to the joint appear thicker on the X-rays. The whole point is that the skeletal system is dynamic. This is to say that the skeletal system is alive and able to compensate for varying loads placed on the skeletal system. Increasing the loads equals increasing the density/mass of the skeletal system. Decreasing the loads equals decreasing density/mass of the skeletal system. It's a very simple concept. This doesn't happen overnight, but over a period of months and years.

How does gravity factor into the mass/density of the skeletal system? Both someone's mass and gravity put axial loads on the skeletal system. The brain is unable to determine if the extra strain on its skeletal system is via gravity or via weight gain. It doesn't matter. The brain will still adjust to both changes in weight and gravity equally by adding bone density.

How do we know this? We know it because of the astronauts losing bone mass from being in space. Their body mass didn't change from being on earth or being in space. So what changed? Gravity was the only factor that changed. And the brain responded proportionately as expected with decreasing bone strength/mass (more porous). Why does the brain do this? It is because the brain involuntarily perceives that the body doesn't need the bone mass like it used to. The DNA coding of the brain is built to be as efficient with energy and material as possible for survival. And when the brain perceives that the extra bone mass is no longer needed, it directs the necessary mineral deposits to other locations in the body.

The whole point is to say that the body adapts to its environment. We see adaptation today with the skeletal system. Dinosaurs are not here to tell us that gravity was weaker some 6,000 years ago, but their skeletal system is here, and it does tell us that gravity was weaker. Unfortunately, paleontologists don't know the skeletal system. They see porous dinosaur bones and porous bones in birds and say, "Hey, there are no more dinosaurs because they became birds." That really is silly. The most likely connection is that dinosaurs adapted to gravity and oxygen changes and grew smaller in size over time.

Review: The body adapts to its environment; specifically, changes in weight bearing on the skeletal system affect the quality/density of the bones. There is a direct link between reduced load on bones and the loss of bone mass (more porous). And there is a direct link between increased load on bones and increased bone mass. Similarly, there is a link between dinosaurs having porous bone structures, with a weaker gravity. The skeletal system responds to changes in gravity and changes in body mass equally. The brain is unable to discern between the two.

We have established the link with gravity and the skeletal system, so how did humans who once lived 900+ years start living less than 100 years? There are multiple contributing factors.

The primary answer is a reduction in oxygen concentration. The advent of the global flood of Gen. 7 caused massive oceans that physically limit vegetation from growing.

Today's current O2 concentration in the atmosphere is 21%. The concentration of O2 before the Flood was ~31%. The oxygen concentration before the Flood was 50% higher than today's value. Today the earth is 70% covered by water, 10% by deserts, and 10% ice, and 10% is usable land that can produce vegetation and O2. Potentially before the Flood, the earth was 30% seas (includes lakes), 0% deserts, 0% ice, and 70% usable land. With an increased concentration of 02 in the atmosphere, living beings get more O2 per breath with each tidal volume of air breathed in. Therefore, dinosaurs could have nostrils that were small in relation to their large size and still get enough oxygen. Why some dinosaurs had small nostrils compared to the volume of air they needed is a puzzle for some, but not for those scientists who are aware that O2 concentration was higher in the past.

The second way that life expectancy was reduced was via increased gravity. The mass of the earth determines gravity. And yes, the mass of the earth is increasing each year. But the rate of mass increase is too small to yield enough change in gravity in a relatively short amount of time, so increased mass is not the means of the increased gravity. And the moon slows the spin of the earth, and this indirectly increases the net effect of earth's gravity. However, that wasn't the primary means either. The means by which gravity was substantially increased was through the removal of a counterforce that was reducing the effect of gravity before the Flood. And after the Flood, this countering force was gone. What is this force? It is buoyancy. The canopy of salt water that hovered around the atmosphere from creation till the Flood put increased atmospheric weight on the inhabitants on earth. This increased weight had a buoyancy effect, which reduced the net effect of gravity.

Reduced gravity allowed blood to circulate through the body with greater ease. And the extra atmospheric pressure allowed increased volumes of oxygen to be breathed in with greater ease with each breath. Imagine an athlete who takes in a greater volume of oxygen in less time (as in living before the Flood) than an athlete breathing in less oxygen (living after the Flood). It would be equivalent to blood doping and allow an unfair advantage. Blood doping is adding extra blood to the vascular system just prior to a competition. The extra blood will carry more oxygen to the muscles and remove more lactic acid from the muscles. This increases endurance. The muscles would have been able to go longer before fatigue set in. Essentially, life thrives in an environment with increased oxygen and reduced gravity.

Also, creatures and humans would be stronger and able to lift objects that would defy our belief today. First, the objects would seem lighter because gravity was weaker. And secondly, the muscles would be stronger because of increased blood flow and increased oxygen. Conversely, with increased gravity, it is harder for life to sustain itself and for the circulatory system to deliver blood throughout the body. With the removal of the buoyancy effect, it's harder to breath in oxygen today than it was prior to the Flood, and less oxygen flows into the lungs with each breath today than at creation.

Review: Reduced oxygen concentration in the atmosphere was the primary contributor to a reduction in the longevity of life. Increased gravity is a second contributor to reducing the longevity of life.

The third way that life expectancy was reduced was via genome adaptation to changes in the environment. The genome is a part of our DNA that we inherit from our parents, and it may be adversely altered by external and internal stimuli. This may affect the body's growth, the number of years of life, and the quality of life. Changes to the environment can involve any external or internal stimuli, ranging from ultraviolet rays, weather changes, diet, sin, and so on.

Here's an example of how a mutated genome may affect future offspring: A male becomes obese and subsequently suffers from diabetes as a result, and he marries a woman that also becomes obese and subsequently suffers from diabetes as a result. They potentially have altered their genomes so that when they have children, they will have a higher propensity to becoming a diabetic than the average child. Why? The parents altered their genome and passed on their altered genes, and this potentially affected their offspring because genomes are the inheritable traits of the parents.

The Bible repeatedly declares that there is a cause and effect in terms of the quality and quantity of our years. For example, Jesus heals a doubled-over paralytic man who had been ailing for 38 years. Then a short time later the same day, Jesus saw the same man in the synagogue and told him to go and sin no more, lest a worse thing come upon him (John 5:1–14), thereby connecting this person's ailment to their sin.

The Bible directly warns people about their sins and indirectly warns about the genome effect on their heirs. God states in the Ten Commandments that He will visit the iniquities of the fathers to the third and fourth generations, while giving blessings to thousands that love Him (Exodus 20:4–6). Medical researchers have recently discovered the ill affect of worrying, anxiousness, and stress, on our body and genome. When someone doesn't trust in God, or know the power of God, they rely on their own devices to handle situations, and this manifests in worry, being anxious, and so forth—which is sin and has an adverse affect on our body. Therefore, the Bible was supernaturally ahead of medical science by 3,000 years, warning parents about the cause and effect of their sins affecting their genomes that are passed on to children. We can either pass down defective genomes resulting from our sin, or we can pass down blessings to our heirs with an obedient life to God. Look at Job; after everything was said and done, God blessed him with additional children: "In all the land no women were found as beautiful as Job's daughters" (Job 42:15). Therefore, our sin may adversely affect our genomes and adversely affect our lives and our heirs' lives. Or our obedience to God will allow the most accurate duplication of the DNA and result in the least amount of mutated genes and benefit us with long lives and benefit our heirs with being beautiful and/or healthy. This is adaptation from internal stimuli from a spiritual realm that affects our physical realm. And this is based on DNA already existing. This is not evolution, where new information in the DNA produces a new function or a new kind.

As mankind continued to sin and live after the fall of Adam and Eve, genomes became increasingly affected as a result. And we see this today with inherited diseases of altered genomes passed down from prior generations. Mankind wants to blame God for this, but it's sin that is the blame.

The best DNA-coded humans were Adam and Eve; as sin continued and generations passed, the copies of the original Adam and Eve's DNA code progressively increased in errors—mutations—which resulted in the decreased efficiency of function or impaired function. This was a result of life adapting to external stimuli from information already existing in the DNA code. The degrading of the DNA code from mutations never enhances/improves function nor results in new kinds of creatures.

Another type of adaptation is modification to a feature that is better suited to environmental conditions. For example, take the Galapagos finches; their beaks seemed adapted to, let's say, reach seeds better, but they already had the information in the DNA code for the differently shaped beak. There was no new information added to the DNA code for a new shaped beak; no, this was from existing information in the DNA code from birth. Finches with the dominate gene trait of a longer beak that better reached the hidden seeds, they thrived because of more food. Thus, when it came time for selecting a mate for reproduction, the mate selected—based on existing DNA that governed the choice to choose the best genetic copy—was the finch with the dominate gene that best reached the seeds. And their DNA dominated the population—this is adaptation, not evolution. If random mutations to the DNA code caused new information that produced a new function, then that would be evolution.

Review: Genome adaptation is the third contributor to reducing the longevity of life. Our actions to God will determine if we pass down blessings (for example, fewest mutations in our DNA and the most accurate copy of our DNA) or pass down curses (for example, increased mutations in our DNA and a more inaccurate copy of our original DNA) to our heirs via our genome, which in turn affects our heirs' lives. For the purposes of this book, the genomes that were passed down from each successive generation post-Flood had built in less and less longevity of life due to the adaptation of environmental conditions (such as increased harmful solar radiation, decreased oxygen, and increased gravity), all of which were a result of God judging mankind's sins.

Evolutionists say there are such similarities in the DNA code from one kind to another kind that it proves we evolved from one common ancestor. Well, this sounds fine, but it equally means we have one common creator. Just like a car, plane, or boat may have similarities in terms of metal, wiring, bolts, and so on, it doesn't mean they came from the same manufacturing plant, but that humans created them. Chimpanzees having similar strands of genetic coding as humans, at first glance, seems to support the notion that humans evolved from primates. But with closer inspection, humans have similar strands of genetic coding to many creatures, not just chimpanzees and not just primates. Why? the genetic code instructs the body to produce proteins that have specific, exclusive purposes. There are similar functions that certain creatures share, such as respiration, reproduction, digestion, taste, smell, touch, CO_2 expiration, waste removal, and so forth. Different creatures also have similar mammary glands for milk, skin, nails, teeth, blood, liver, hormones, hair, eyes, ears, muscles, tendons, ligaments, adipose, bone structure, and so on. Of course there are similarities in the genetic code for building 3D proteins that have similar specific, exclusive functions. Since there is a specific genetic code to build specific proteins that make tendons that attach muscle to bone, it makes perfect sense that the genetic code would be similar for all creatures that have the same ligament composition. Since there is a specific genetic code to produce blood from the marrow of the trabeculae in bones, well of course the genetic code would be similar for all creatures that produce blood from the marrow of bone. The inference is that we share similar DNA because we share similar proteins that perform similar functions. There is no mystery here; this is not evolution. This is God being efficient with DNA that performs the same function in different creatures—just as a computer programer, utilizing the same operating system, wouldn't write a new and different program that performs the exact same function of an already existing program. The software writer would simply copy and paste the functional information.

Therefore, similarities in the genetic code do not demonstrate that we evolved from primates. If so, we would not have similar strands of genetic coding with other creatures on a completely different evolutionary branch. What similarities in the genetic code demonstrate is that there was one creator that created different kinds from the same dirt. It is common to hear evolutionists purport that these similarities are proof of evolution, but the logic bypasses the fact that humans have genetic similarities with other creatures as well—even with bananas. The Bible says that God made everything through Himself, so of course there will be similarities in genetic code.

Evolutionists look at general similarities, such as the fact that one animal has five bones in a limb and another has five bones in a limb and draw the conclusion that they came from the same ancestor. But a basic genetic factor of linking one kind of creature to another is the chromosome count. But when studying the chromosomal count of each kind and each species within the kind, there is an apparent widespread variety in chromosomal count. For example, Chorda algae have 56, Cosmarium algae have 120, Bacillus fungi have 1, Saccaromyces fungi have 30, Amoeba protozoa have 30, Radiolaria protozoa have 800+, and on and on. Let me simplify my train of thought. Humans have 46 chromosomes, and antelopes have 46 chromosomes and lots of similarities. But there are hundreds of creatures with similar chromosome counts (or exactly 46 chromosomes); this doesn't prove or disprove anything. Evolutionists teach that life evolves and becomes more complex, but this is contrary to humans having 46 chromosomes and white ash trees having 138 chromosomes or ferns having 480 chromosomes, as they preceded humans on the evolutionary tree. It seems more like entropy (a law that says that everything tends from order to disorder) rather than evolution.

Creationists and evolutionists are looking at the same evidence, but the condition of the soul and spirit determines the interpretation. If one determines that there is no God, then all interpretations will be anti-Bible, anti-creation, and anti-God. If one determines there is a God from the supernatural work of the Bible, then the interpretations will be pro-Bible and pro-God.

Review: Similarities in genetic code point toward one creator. Creatures that have similar functioning things, such as tendons, blood, and muscles, have similar genetic codes. The dissimilarities in chromosome counts are contrary to the concept of evolution.

After the Flood, more solar radiation reached the earth's surface, and oxygen concentrations in the atmosphere gradually declined as net gravity increased. They had adverse effects on the duration and size of life. This resulted in the genomes of all living organisms adapting to the new environment. This adaptation of the genome meant that each subsequent offspring would not live as long and grow as tall and as massive. The genes being passed down to the next generation were affected with each passing generation until this process leveled out to the homeostasis we have seen in the last ±3,000 years.

Yes, this means dinosaurs would have been affected as well. With each passing generation, they too would have been smaller and smaller in size. Dinosaurs are reptiles, and they grow as long as they are alive. So if their life expectancy was cut shorter and shorter with each passing generation and gravity was greater after the Flood, then the emphasized genome that was passed down had information that allowed the once mammoth sizes to adapt. Yet, all the creatures on earth still have in their DNA code the necessary information to once again live 900+ years.

  We know dinosaurs survived the Flood. How do we know this? God ordained that two of every kind of creature on earth, male and female, would enter into the ark (Gen. 6:19–20, 7:8–9, and 13–17). Noah didn't physically gather every kind of creature in the Ark; that would have been impossible. Some evolutionists resort to twisting what the Bible records, creating a "straw-man" argument to discredit the Bible. For example, they suggest that it is impossible that Noah got every creature on the tiny boat. The Bible records that two of every kind went on the ark, not two of every species; there are hundreds of species in a kind. For example, there weren't thousands of dog species on the ark; there was one male wolf and one female wolf.

  I have heard countless times from evolutionists: "It's physically impossible for <u>Noah to</u> gather two of every kind into the Ark. Therefore, the Bible is wrong." But this is a silly argument. For the text clearly states, in Gen. 6:18–22, "You shall go into the ark . . . <u>two of every kind **will come to you**</u> to keep them alive." Also, Gen. 7:7–12 says, "So Noah, with his sons, his wife, and his sons' wives, went into the ark . . . clean animals and unclean animals, and birds, and everything that creeps on the earth, <u>two by two **they went** into the ark **to** Noah . . . **as God commanded**</u>." And again in Gen. 7:14–16: "<u>They [the animals, two by two] went into the ark to Noah</u>, two by two, of all flesh in which is the breath of life. So those that entered, male and female of all flesh, <u>went in as God had commanded him</u>; and the LORD shut him in."

  Of course, the dinosaurs that went into the ark were younglings. This seems too obvious to point out. But evolutionists use this ploy to try in vane to invalidate the Bible by saying, "It's physically impossible to get full-grown dinosaurs in the ark." But when they are young, they eat less, poop less, and take up less space. Therefore, it's logical that most of the larger creatures would have been young.

  Do we have additional proof that dinosaurs survived the Flood? We don't need more than the Bible, but let's consider the book of Job. Job lived approximately six generations after the Flood. Yet, the book of Job describes dinosaurs in Job chapters 40 and 41. In Job 40:17, God talks about a behemoth with a tail like a cedar tree. Cedar trees in the Mesopotamian era grew 75–100 feet tall. That can only refer to a dinosaur. God also talks about the leviathan that He created. This leviathan had fire coming out of his mouth. The whole of the chapter can only refer to a dinosaur. Why? Consider the bird's-eye view of the chapters. God is saying that He is so much greater than any man because man cannot tame or conquer the greatest beasts that God created. Logically, the chapters of Job 40 and 41 refer to dinosaurs. God would be making a silly argument if He is saying that Job so inferior that he can't conquer the hippopotamus. Also, It is a worthless argument to reference something that the hearer doesn't know about to solidify ones point. Therefore, Job was fully aware of these dinosaurs, otherwise, God saying, "Behold" ("look at") is pointless in referencing these creature to Job.

  I have no problem taking these chapters literally—even the fire-breathing dragon. Why? Today we see the bombardier beetle as a small version of how this may have played out. The bombardier beetle has three separate fluid-filled chambers in its abdomen. When the bombardier beetle gets scared, it sends two of the fluid-filled chambers to drain into a distal empty holding tank; once the two fluids are mixed, then a third reservoir that is filled with the catalyst enzyme drains into the holding tank. With the combination of these three different chemicals, the fluid starts to heat up to 200° F and boil under pressure, and then when the pressure exceeds the resistance threshold of its rear sphincter, a noxious boiling fluid bursts through the rear sphincter when a predator stalks the beetle. All this is done in fractions of seconds. The scalding hot noxious chemicals burn the unsuspecting predator with a chemical burn and temperature burn. And while smoking hot, the predator flees. We see today how this could have been possible with a fire-breathing dragon (literally).

  What's significant about this? God is talking about dinosaurs to Job, and for Job to understand God's point, God discussed something Job could understand and relate to. Therefore, Job knew what God was talking about. This is Biblical evidence that dinosaurs existed after the Flood and corroborates the Genesis account, which says that all creatures, including dinosaurs, went into the ark.

Therefore, dinosaurs had to go through the same adaptation as any other creature that lived in the post-Flood era, including humans. This suggests that as humans had to adapt and their longevity was reduced, then so too the dinosaurs and all life adapted to less and less oxygen and adapted to increased gravity. They would have lived shorter and shorter lives. Since dinosaurs are reptiles, and we know reptiles continue to grow as long as they are alive, then it makes perfect sense that as the dinosaurs lived shorter lives, they would have grown to shorter and smaller maximum sizes with each successive generation.

In fact, dinosaur fossil footprints have been found on the same layer of soil as human beings. This means that they both walked on the face of the earth during the same time frame. This evidence supports the Bible and a plain interpretation of the Genesis creation account and the Flood account. Where are these footprints found? They are in the Paluxy River near Glen Rose, Texas. There isn't just one human footprint or just one dinosaur print, but many of each. The average length of the human footprints is 11.5 inches in length, with alternating steps that are among and within dinosaur tracks.

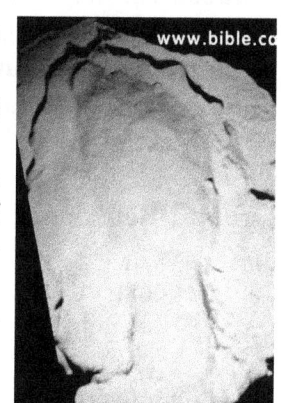

Above, you can see dinosaur footprints going one direction and a human's going another. The picture on the right is a plaster cast made of one of the footprints. The findings of dinosaur prints and human prints are in harmony with the Bible but a problem for evolutionists. For there are ~65 million years between dinosaurs and human existence, according to the evolutionary hypothesis. But the Bible has them both living at the same time.

Review: The Bible is in harmony with dinosaurs living in the post-Flood era. Gen. 7 declares that one male and one female of every kind of creature entered the ark. And in Job 40 and 41, God tells Job to look at two dinosaurs that He made. God is talking about dinosaurs, as the leviathan had a tail as long as a cedar tree (75–100 feet), and the behemoth was a fire-breathing dragon. *Photo credit: www.bible.ca by Stan Taylor.*

Before the global flood of Gen. 7, gravity was weaker. This would have had a profound effect on all skeletal systems of all creatures living in that environment. Their skeletal systems would have been more porous and had less mass/density than if they lived today.

Since dinosaurs had porous bone structure relative to their massive tonnage, this is evidence that dinosaurs lived in a gravitational environment that was weaker and explains why their bones were porous, which sounds much more plausible than the notion that they evolved into birds. By taking a simpler interpretation of the evidence, evolutionists don't have to go to great lengths to establish their belief, such as falsifying evidence to support such a leap of faith. I'm speaking about Archaeoraptor.

The Bible records humans living 900+ years before the Flood, which means all life, not just humans, lived a very long time. With reptiles growing as long as they are alive, and living 900 years in an environment with weaker gravity, increased oxygen, less harmful solar radiation, and ambient temperatures; it's not hard to imagine that they would be huge, and we would call them dragons or dinosaurs. The Bible gives us the clue that humans lived so long, so we can infer that animals did too. And this explains how we got dinosaurs/dragons on earth. Another interesting point is that humans have primary growth sites that fuse early, ending vertical growth, but secondary characteristic sites can continue to grow. Secondary characteristic sites include the brow above the eyes, cheek bones, and jawbones, and so on. In fact, facial bones never stop developing. When evolutionists find skulls with slightly pronounce brow ridges, this doesn't conflict with human anatomy or the Bible. Imagine primates living ~10 times today's longevity, and humans living 900+ years and both being taller and having more pronounced brow ridges, cheek bones, and jaw bones, as they approached almost a thousand years of age. This seems to be what all the evolutionists are claiming is the missing link with humans and primates. Evolutionists attach names such as hominids, Cro-Magnon, and so on to these old skeletal fragments, but they could very well be from humans and primates that had pronounced secondary characteristic sites due to prolonged life.

Chapter Summary: The evidence is that dinosaurs have porous bone structures. Given that bones respond to changes in loads and that the canopy of water hovered around the atmosphere to reduce gravity, it is conceivable that dinosaurs lived in a weaker gravitational system along with humans. Living things once lived 900+ years because oxygen was 50% higher and gravity was weaker. The global flood changed all that. The canopy rained upon the earth, which eventually reduced the oxygen concentration and removed the buoyancy effect. As the genomes of DNA adapted to the changes in the environment, the maximum life spans declined with each successive generation. Dinosaurs are reptiles, and reptiles grow as long as they are alive, so when the dinosaur life expectancy declined, so too did their size. And when gravity increased, their skeletal structures adapted. Since dinosaurs survived the Flood, what happened to them?

Group Discussion:

1. This chapter presented a plausible explanation for how the patriarchal ages gradually reduced through the generations following the Flood. How did this affect your faith as you gain in knowledge?

2. Knowing potentially how God caused the ages to reduce, does this hint to you how God might reverse the curse on nature and cause mankind to once again live 900+ years?

Chapter 25
What Happened to Dinosaurs?

Did all the dinosaurs die off millions of years ago? Well, the Bible describes that one male and one female of every creature went into the ark so that would include dinosaurs (Gen. 6). However, they would have been very young dinosaurs to accommodate for size, food consumption, and aggressiveness. Therefore, we may infer the Bible declares that dinosaurs survived the Flood. Right off the bat, those who claim to believe in God are faced with a dilemma. Do you believe the wisdom of men telling you that dinosaurs died off hundreds of millions of years ago, or do you believe in the inerrant Word of God that declares one male and one female of every creature went into the ark?

Now that we know the Bible indirectly tells us that dinosaurs survived the Flood, do we know that the dinosaurs lived long after the Flood? Yes, Gen. 6:19: "And of every living thing of all flesh, you shall bring two of every kind into the ark, to keep them alive with you." Therefore, **dinosaurs lived after the Flood because God decreed "to keep them alive with you."**

Furthermore, God talks about dinosaurs as being alive and in the present tense to Job to prove His point. God uses the words *leviathan* and *behemoth*, but through time and dialect changes, humans called these creatures dragons and then dinosaurs. God makes a point to Job by using an illustration that is only valid if the listener is aware of His example. Therefore, Job would have been familiar with the leviathan and behemoth (Job 40 and 41). When did the conversation between God and Job occur? There is no direct date given, but there are enough clues provided in the Scriptures that we may discern an approximate date. Job 22:15–16 (NIV) indicates that Job lived post-Flood, "Will you keep to the old path that the wicked have trod? They were carried off before their time, their foundations were washed away by a flood." Also, Job lived during the time of the patriarchs because as the head of his family, he offered up sacrifices to God (Job 1:5). Job 42:8 indicates that the friends of Job also offered up sacrifices to the Lord, a practice that ceased during Moses' time with the law given by God, and the priests performed the sacrifices on behalf of the people. And Job gave inheritance to his daughters (and sons), a practice that was only done in the absence of brothers to preserve the twelve tribes (Numbers 27:1–11). And Job's wealth was valued in the amount of his livestock, not in money (Job 1:3, 42:12). And finally, the strongest bit of evidence is the length of years Job lived. Job 1:13–18 suggests that Job lived long enough to accumulate great wealth and have children old enough to be at the age of accountability. And Job 42:10–17 reveals that after everything was taken from Job, he lived an additional 140 years. Thus, putting the pieces together, Job lived 220+ years, about as long as Peleg (239 years), Reu (239 years), and Serug (230 years), around six generations after the Flood. Therefore, the conversation between God and Job occurred around the time of ~2,200 BC.

Let's take a closer look at the two chapters in Job that talk about dinosaurs to confirm that these creatures being described are not an elephant, hippopotamus, or whale. Let's set the scene: Job was a righteous man that God had blessed so much that he had eight children and honor, and he was the richest man of the East. God was so pleased with Job's obedience that God allows Satan to test Job and take away Job's children, his health, and his riches to demonstrate to Satan that Job loved God regardless of the blessings that God had given Job.

Satan kills Job's children, causes thieves to steal all his wealth, and afflicts Job's body with a host of sicknesses. Job is broken to the point that he questions God. Then God answered Job out of the whirlwind and said, (Job 38:1–3) "Who is this that darkens counsel by words without knowledge? Now gird up your loins like a man, and I will ask you and you instruct Me!" God asks Job a series of questions, all of which Job is unable to answer. God is establishing that His ways are above our ways. And God doesn't make mistakes ever, let alone with His plan or justice.

Here are some of the questions God asks Job: Job 38:4 "Where were you when I laid the foundation of the earth? [If you are as mighty as God, why weren't you there when I formed the earth?]

. . . Who set its measurements? . . . who stretched the line on it? On what were its bases sunk? Or who laid its cornerstone, when the morning stars sang together and all the sons of God shouted for joy?" [On the third day of creation].

In verse 8, in reference to the global flood when the waters were bursting out of the earth, God asks, "Or who enclosed the sea with doors when, bursting forth, it went out from the womb; When I made a cloud its garment and thick darkness its swaddling band, and I placed boundaries on it and set a bolt and doors, and I said, 'Thus far you shall come, but no farther; and here shall your proud waves stop?'" [Job wasn't there when the Flood occurred and can't respond to the inquiry].

In verse 12, God asks, "Have you ever in your life commanded the morning, caused the dawn to know its place, that it might take hold of the ends of the earth, and the wicked be shaken out of it?" [Job wasn't there at creation and therefore can't respond to an all powerful God].

In verse 15, He asks, "Have you entered into the springs of the sea or walked in the recesses of the deep? Have the gates of death been revealed to you or have you seen the gates of deep darkness?" [Job wasn't there at the creation of the seas on the third day of creation, nor has he seen Hell, so he cannot answer any inquisition].

In verse 18, He asks, "Have you understood the expanse of the earth? Tell Me, if you know all this." [On the second day of creation, Job was not there to see how the atmosphere was created].

In verse 19, He asks, "Where is the way to the dwelling of light? and darkness, where is its place, that you may take it to its territory and that you may discern the paths to its home? You know, for you were born then, and the number of your days is great!" God is making a clear point to Job that he doesn't know much because Job wasn't there when light was created. Job doesn't know the answers.

In verse 22, He ask, "Have you entered the storehouses of the hail, Which I have reserved for the time of distress, For the day of war and battle?" This is hail so severe that it's judgment hail, not ordinary thunderstorm hail (Revelation 16:21: huge hailstones, weighing about 100 lb.).

In verse 25, He asks, "Who has cleft a channel for the flood, or a way for the thunderbolt, to bring rain on a land without people, on a desert without a man in it, to satisfy the waste and desolate land and to make the seeds of grass to sprout? Has the rain a father? or who has begotten the drops of dew?" [Humanity does not know the path of the thunderbolt; nor did Job.]

In verse 31, He asks, "Can you bind the chains of the Pleiades, or loose the cords of Orion? Can you lead forth a constellation in its season, and guide the Bear with her satellites? Do you know the ordinances of the heavens, Or fix their rule over the earth?" Pleiades is a constellation that humanity cannot bind with chains. Humanity also can't release the gravitational grip that Orion has.

In verse 34, He asks, "Can you lift up your voice to the clouds, so that an abundance of water will cover you? Can you send forth lightnings that they may go and say to you 'Here we are?'" Job didn't control the rain and hydrological cycle; nor do we. We command not where the lightning goes or when it comes.

In verse 36, he asks, "Who has put wisdom in the innermost being or given understanding to the mind? Who can count the clouds by wisdom, Or tip the water jars of the heavens, When the dust hardens into a mass and the clods stick together?" Job and all believers know that the wisdom of mankind comes exclusively from God; this is from James 1:5 and Colossians 2:2–3. Therefore, Almighty God is mightier than the created.

In verse 39, He asks, "Can you hunt the prey for the lion, or satisfy the appetite of the young lions, When they crouch in their dens and lie in wait in their lair? Who prepares for the raven its nourishment when its young cry to God and wander about without food?" Job knew he did not prepare food for all of creation each day. Colossians 1:17 reveals that Jesus is the sustainer of His creation.

Job 40: In verse 1, He asks, "Will the faultfinder contend with the Almighty?"

Verse 3: "Then Job answered the LORD and said, 'Behold, I am insignificant; what can I reply to You? I lay my hand on my mouth.'" Job gets the point that he is too insignificant to question God.

But God is not done making His point. In verse 6, He says, "Then the LORD answered Job out of the storm and said, 'Now gird up your loins like a man; I will ask you and you instruct Me.'"

In verse 8, He asks, "Will you really annul My judgment? Will you contend with Me that you may be justified? Or do you have an arm like God, And can you thunder with a voice like His? Can you adorn yourself with eminence and dignity . . . and look on everyone who is proud, and make him low. And Look on everyone who is proud, and humble him, and tread down the wicked where they stand?"

God describes His greatest land creature, the behemoth in verses 15–17: "Behold now, Behemoth . . . He bends his tail like a Cedar." Cedar trees that grew in the Mesopotamian era and region grew to 100 feet tall. This hardly fits the hippopotamus, with its very short tail, and it rules out the elephant. God wouldn't use all these grandiose illustrations to then use the tiny tails of the hippopotamus and elephant. That is incongruous with all His prior examples.

In verse 19, He says, "He is the first of the ways of God." As far as the land creatures go, this animal is the most impressive. It was the first as in terms of strength, not order, since all land creatures were created on the sixth day (Gen. 1:25, "great beast").

In verse 24, He asks, "Can anyone capture him when he is on watch, with barbs can anyone pierce his nose?" This doesn't seem to fit elephants either, as man has tamed elephants. And elephants have a trunk instead of a nose. Dinosaurs fit all the categories of the description in Job 40:15–24. The hippopotamus and elephant fit some of the descriptions, but not all. What animal would be the first of the ways of God? The great sauropod, Amphicoelias fragillimus, was 60 meters in length, (200 ft.), weighing in at a whopping 150 tons. This is the largest relative of bronchiosaurus.

Review: Job 40:15–24 indicates that the behemoth was a type of dinosaur. And it seems congruent that God is describing one epic portion of His creation after another, one mighty thing after another when He then describes something like bronchiosaurus. *Image credit: wikipedia.com*

The leviathan described by God is even more impressive than the behemoth. In Job 41, verse 1, He asks, "Can you draw out Leviathan with a fishhook? Or press down his tongue with a cord? Can you put a rope in his nose or pierce his jaw with a hook?" Mankind has captured every creature in the sea with some type of bait, hook, or net, but not a fire-breathing dinosaur. Therefore, this rules out the blue whale. Right off the bat, in the first verse, God is talking about something so monumental that the blue whale doesn't fit. This is in parallel with all the prior illustrations.

In verse 3, He asks, "Will he make many supplications to you, or will he speak to you soft words? Will he make a covenant with you? Will you take him for a servant forever? Will you play with him as a bird, or will you bind him for your maidens?" No dinosaur made a request to Job or spoke softly to him, and no human played with a dinosaur. Yet, we have many sea creatures in zoos for our pleasure. Even the killer whale is bound for our amusements.

In verse 6, He asks, "Will the traders bargain over him? Will they divide him among the merchants? Can you fill his skin with harpoons, or his head with fishing spears?" Again, this rules out the mightiest ocean dwellers we see today. For mankind has traded and divided the spoils of whale blubber for a millennium. But no man has ever captured or tamed Livyatan melvillei, with a head 10-foot long and teeth that were over 1-foot long. This fierce sea creature was like a killer whale but the size of a sperm whale. Or how about Megalodon? This was a remarkable prehistoric shark on steroids—the size of a school bus in length.

In verse 8, He says, "Lay your hand on him; Remember the battle; you will not do it again! Behold, your expectation is false; Will you be laid low even at the sight of him? No one is so fierce that he dares to arouse him; Who then is he that can stand before Me?" And there lies the point of God describing the greatest of sea creatures, rather than a whale. Only the prehistoric dinosaurs make man realize his insignificance versus the created, let alone versus the creator.

God goes into details about this great sea creature that reveals clearly that this animal was unlike anything we see today, but alive and well in Job's time. In verse 13, He asks, "Who can open the doors of his face? Around his teeth there is terror. His strong scales are his pride, shut up as with a tight seal. One is so near to another that no air can come between them. They are joined one to another; They clasp each other and cannot be separated."

The scales rule out large aquatic life such as whales and dolphins, and fits a reptilian aquatic dinosaur. The terror around his teeth rules out the large aquatic creatures with baleen in their mouth. What kind of terror? In verse 18, He continues,

> His sneezes flash forth light, and his eyes are like the eyelids of the morning. Out of his mouth go burning torches; Sparks of fire leap forth. Out of his nostrils smoke goes forth as from a boiling pot and burning rushes. His breath kindles coals, and a flame goes forth from his mouth.

Clearly, this is unlike all creatures on earth today. We see today that a fire-breathing creature, as described in Job 41, is within the realm of possibility. In chemistry class, students learn that certain liquid chemicals create fire without matches or a lighter when mixed together. In the following examples, only drops of chemicals are needed.

Chemical fire combination #1: Potassium permanganate + glycerin + water = fire. This is a nice fit since water is required. Remember that this particular leviathan that God is describing is in water yet sends fire out of its mouth. Potassium permanganate is a salt containing manganese and potassium; a mixture of manganese dioxide and potassium carbonate. And glycerin is a sugar alcohol. A living organism, with the appropriate proteins, would be able to process ingested food and fill separate chambers of potassium permanganate and glycerin.
Chemical fire combination #2: Acetone + sulfuric acid + potassium permanganate = fire.
Chemical fire combination #3: Sodium chlorate + sugar + sulfuric acid = fire.
Chemical fire combination #4: Ammonium nitrate powder + zinc powder + hydrochloric acid = fire.

Who knows what method God used to form this dinosauric fire-breathing sea dragon. Before you think this implausible, remember that the bombardier beetle has a chamber filled with hydroquinone and another chamber filled with hydrogen peroxide. If you think it's impossible to house noxious chemicals inside a living organism, consider that humans have a chamber (stomach) of fluid in their body that holds hydrochloric acid. This is one of the deadlier known acids to man. If this acid comes in contact with any other body part, it can result in severe injury or death. Everyone has it in their stomach, and so do many other animals. Also, snakes have sacks filled with toxic venom in their skull without suffering any ill effects. Therefore, a fire-breathing leviathan is plausible. Matthew 20:26 says, with man some things are impossible, "but with God all things are possible.'".

Mankind has not yet identified a dinosaur fossil that matches the Biblical description of the fire-breathing dragon. However, if the bombardier beetle were extinct and discovered as a fossil, would we discern that the three chambers in the beetle's abdomen were used to generate a 100°C noxious gas that chemically and thermally burns a nearby predator in a fraction of a second? Probably not. Without modern-day chemistry, would we have known that potassium permanganate + glycerin + water = fire? Who would have thought that water was a requirement for a fire?

Therefore, don't shy away from taking these verses on face value. For the other verses and chapters are uniformly interpreted on face value. Just because the words describe a creature that we don't see or know of does not mean that they are not in harmony with the surrounding verses. Jumping to a figurative view may be appropriate at times, such as when the Bible says behemoth had legs like iron or a heart as hard as stone. It is not literally saying the legs were iron or that the heart was stone. But to default to a figurative interpretation of the Scriptures because a literal view causes one to be a fool in the eyes of others, is not justification to do so. And it may reveal a faith that is weak regarding the abilities of God and the infallibility of His Word. It's not as though the writer of the book of Job, even though inspired by God, got a little carried away with the pen at this point. When in doubt on how to interpret the Scriptures, default to a literal interpretation. This will save you a lot of contradictions and troubles later on. Continuing with verse 25: "When he raises himself up the mighty fear; Because of the crashing they are bewildered. The sword that reaches him cannot avail, nor the spear, the dart or the javelin. He regards iron as straw, bronze as rotten wood. The arrow cannot make him flee; Slingstones are turned into stubble for him. Clubs are regarded as stubble; He laughs at the rattling of the javelin. His underparts are like sharp potsherds."

This language describes no creature alive today, but it fits dinosaurs. Just as some dinosaurs had protruding and jagged plates of scales, so too did this dinosaur. Verse 31:

He makes the depths boil like a pot; He makes the sea like a jar of ointment. Behind him he makes a wake to shine; One would think the deep to be gray-haired (mane). Nothing on earth is like him, one made without fear. He looks on everything that is high; He is king over all the sons of pride.

The heat from the chemical fire proceeding from his mouth causes the water to boil, and the oily mineral this dragon uses to create fire leaves an oily residue floating in the water behind him.

Review: Fire-breathing sea creatures are possible. And the writer of Job uses present tense grammar. Therefore, Job was able to see this leviathan and understand what God was talking about because dinosaurs lived after the Flood.

Therefore, the Bible is clear that dinosaurs not only lived with mankind, but survived the Flood. If dinosaurs survived the Flood, then where are they? When and how did they all go extinct?

We know that life thrives in an oxygen-rich environment with a reduced gravitational force, and conversely, we know that life suffers in an oxygen-poor environment with an increased gravitational force. Perhaps some of the dinosaurs are extinct now, but perhaps many of them still live among us in an adapted state. Humans had to adapt to the changed conditions after the Flood; so too did all life have to adapt, and this includes dinosaurs.

For example: Take the saber-toothed tiger. It was much larger than tigers today, but what would happen if a tiger lived in a high-oxygen and weak-gravity environment for many generations? The tiger would live longer and grow taller. The teeth would also be larger. In fact, I contend that the canine tooth that distinguishes the saber-toothed tiger from other large cats would grow just as long on the tiger. And conversely, if the Saber-toothed tiger lived today, its elongated canine tooth that makes it so distinguishable from other large cats would be reduced to the same size as the tiger today.

Therefore, it's possible for the saber-toothed tiger to be the forefather of the current tiger. And the only difference between the two is the environment they live(d) in. Just as our forefathers once lived 900+ years, so too did the tigers' forefathers once live, with long canine teeth that were a result of the environment and not because of evolutionary change. They merely adapted to changes in the environment. This means that if oxygen levels and gravitational forces return to their pre-Flood optimal levels, so too would the tiger's canine tooth elongate over time to the length of its ancestors, the saber-toothed tiger.

Review: The saber-toothed tiger could plausibly be the ancestor of the modern-day tiger, which means the saber-toothed tiger is not extinct, but living amongst us as the tiger.

What about the mighty Tyrannosaurus rex? Did the T. rex become extinct as well? When oxygen levels gradually declined with each passing decade after the Flood, T. rex would have gradually grown to a lesser height with each subsequent offspring, as it adapted to the changes in oxygen. Remember, in an oxygen-rich atmosphere, life thrives. But in an oxygen-poor atmosphere, life suffers. Therefore, the T. rex would gradually be reduced in size, strength, tonnage, and longevity of life. Since T. rex was a reptile, if we cut short its life span, we also cut short its size.

Now let's increase the gravitational force and see what effect that has on T. rex. Remember the example of the astronaut losing 14% bone mass because of the weightlessness of space. Okay, back to T. rex. With the increase in net gravity after the Flood, there was an immediate change. Though T. rex was very strong and powerful and may have been able to resist this change for a generation of two, it was unable to endure the increased gravity and remain upright as a bipedal creature, so on its belly it went, adapting with modified means of survival and mobility. T. rex learned that mobility was easier in the water. Those that made it to the watery habitats survived and learned a new way of life. Being on his belly forced its head and neck up. It still possessed a large powerful tail, but this large powerful tail was no longer used for erect posture and balance, but was now used for mobility in the water. Does this sound like any creature we may know and see today? How about the alligator? Don't they look like a T. rex that has adapted to two major changes in environment, reduced oxygen, and increased gravity?

It seems quite plausible that T. rex is not extinct, but living amongst us as the alligator. And likewise, the velociraptor, a smaller relative to T. rex, may also be living amongst us as the crocodile. The crocodiles have smaller jaws than the alligator, and velociraptor had smaller jaws than T. rex. Otherwise, they had similar characteristics. *Photo credit: Ricky Flynt/Mississippi Wildlife, Fisheries, and Parks Dept.. Image credit: www.bbc.co.uk/nature/life/Tyrannosaurus. by John Sibbick.*

With the plausibility that the T. rex adapted to become the alligator and velociraptor adapted to become the crocodile, what are some of the similar characteristics? All four lay a leathery egg. They are vicious predators at the top of their ecosystem food chains and have short and weaker upper/forward arms, stronger and larger hind legs, a long strong tail, an elongated snout, eyes perched on top of heads, and reptilian skin with ridges on their back and smooth underbelly. They also have a proportioned tail-to-body length; the list keeps going. Just stand back and take a bird's-eye view of both the T. rex and the alligator and the velociraptor and the crocodile. The similarities are awesome. If we were able to travel back in time and see how increased gravity and decreased oxygen affected T. rex and velociraptor, we would expect to see them forced on their bellies and crawling, utilizing the water to mitigate against gravity. They would be smaller in size because of a reduction in oxygen. And we would expect to see them look and act like our modern-day alligator and crocodile.

Review: It's plausible that T. rex is not extinct, but living amongst us as an alligator, and that

velociraptor is not extinct, but living amongst us as the crocodile.

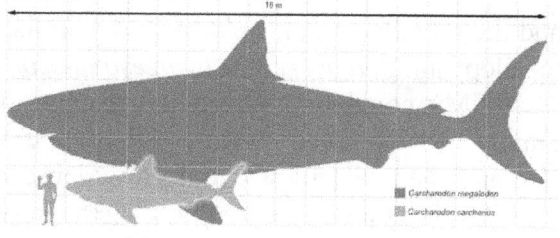

How about the mighty Megalodon? They had teeth 5 inches long, and their bodies were approximately 50 feet long—the size of a school bus, but with jaws. Now reduce the oxygen in the water from pre-Flood conditions of, let's say, 16%, to the current-day concentration of 7%–8% oxygen levels. This would decrease growth rates and affect the size of Megalodon, but it would adapt to produce offspring that were smaller and smaller in size until we got to a size that is similar to our great white shark. Take a bird's-eye and a worm's-eye view of the Megalodon and the great white shark; they have the same features, proportions, and characteristics. It's quite plausible that Megalodon is not extinct, but living amongst us as the great white shark. *Image credit: comicvine.com.*

The great white shark still has the information in the DNA to return to Megalodon size. If we reversed the oxygen concentrations back to pre-Flood conditions and decrease gravity back to pre-Flood conditions, the great white shark would adapt and produce offspring that increasingly got larger and larger, until the size of Megalodon was restored.

Review: It's plausible that Megalodon is not extinct, but living amongst us as the great white shark.

Take the mammoth, which is said to be extinct. But it has all the external and internal characteristics of our modern-day elephant. I contend that if we were living at the time of the mammoth and seeing through an accelerated time of several generations, we would see the mammoth adapting to decreased buoyancy and increased gravity and decreased oxygen in the atmosphere. We would see changes to the following features through the generations: a shorter tusk (but consisting of the same tusk composite), the same short tail, the same trunk (but smaller), and basically the same image, but smaller. They would share the same likeness, personality, and diet. Essentially, if we watched the mammoth adapting to the changes in the environment through a time-lapse film, we would see the mammoth adapt to the elephant over a thousand-year period, not a changing of kind, but merely the same kind adapting.

What about the hair? The mammoth had lots of dense hair, and the elephant seems bald. Well, the elephant is not bald; it has the same type of hair, but it's just not dense. It's like they adapted to the reduced oxygen by almost fully removing their hair. But they still have hair that is sparsely dispersed around their body. Is this so far-fetched? Consider that losing hair is not impossible; some male humans lose their hair as they age. And all evolutionists believe that human beings shed their entire body hair heading into the ice age. Which seems less likely—that human beings shed their body hair heading into the ice age or that mammoths shed their hair from a lack of oxygen and adapted to the new environment as the elephant? Come on; it's foolish to think that through means of random, unguided mutations, humans shed their warming body hair while entering the ice age. We've already linked high oxygen levels with thriving life, and for the mammoth, that thriving life was represented by their large tusk, mammoth size, and dense hair. Elephants still have the tusk, size, and hair; they are just not as developed due to the

environment.

If one takes a bird's-eye view and a worm's-eye view of both the mammoth and the elephant and their similarities, it's plausible to surmise that the mammoth is not extinct, but living amongst us as the elephant. *Image credit: strangesounds.org.*

My contention is that if we grew elephants in an oxygen-enriched atmosphere and reduced gravity and increased buoyancy, they would, based on information in their DNA, increase their tusk size and their overall size and grow dense hair again. This may not occur in one generation, but we would see slight changes with each generation that lived in a pre-Flood environment.

After all, male pattern baldness in men is due to a lack of oxygen supply to the hair follicle. Most people only hear that male pattern baldness is due to a lack of blood supply, but what does the blood carry? Oxygen. With a lack of oxygen, life suffers. But here, we are talking about the mammoth's hair. With a lack of oxygen, the mammoth's dense hair gradually declined in quantity and quality.

Review: It's plausible that the mammoth is not extinct, but living amongst us as the elephant.

Everyone has seen drawings of stegosaurus (picture below), the dinosaur with the spinal plates on its back. Well, there is a reptile that lives in New Zealand called a tuatara that has a distinctive feature that is interesting to visualize. The tuatara has similar-looking, albeit much smaller, spinal plates or exterior spinal crest (fins) as the dinosaur stegosaurus had. Tuatara's spinal plates point away from the center of the spine in an alternating fashion similarly to stegosaurus. The name *tuatara* means peaks on the back.
Photo credit: Wikipedia.org. Photo credit: www.thereptilereport.com. The tuatara has similar skeletal and internal structures to stegosaurus and spinosaurus. It's plausible that either of those two dinosaurs adapted to the increased gravity and reduced oxygen and significantly reduced longevity of life and live today as a much smaller reptile. Any minor variances between the two are explained via the adaptation process over the generations from the pre-Flood conditions (low gravity, high buoyancy, and high oxygen) to the post-Flood (high gravity, low buoyancy, low oxygen) conditions. Spinosaurus seems to have similarities too. *Photo credit: Wikipedia.org. Photo credit: Alsipnursery.com.*

If we started with the tuatara reptile and raised it in a hyperoxic environment, with a high buoyant force and weak gravity, then the tuatara lizard would enlarge over time with each generation, and the spinal plates would enlarge to the size of a pre-Flood ancestor—potentially stegosaurus or spinosaurus.

If we start with the stegosaurus and work forward in time to just after the Flood, then the spinal plates on the back of stegosaurus living in post-Flood conditions (high gravity, low buoyancy, and low oxygen) would gradually get smaller and smaller, and so too would the size of the stegosaurus, until we arrived at the size of the tuatara or something similar. This means that stegosaurus or spinosaurus is potentially not extinct, but possibly living amongst us as the tuatara. The picture is of spinosaurus and a small reptile.

Review: It's plausible that the stegosaurus and spinosaurus are not extinct, but living amongst us as reptiles, possibly as the tuatara or some other lizard.

Evolutionists might critique this approach by noting that creationists don't believe there is enough time for evolution, yet they believe that 4,000 years is enough time for T. rex to evolve into an alligator or for velociraptor to evolve into a crocodile. No, that is not what I'm saying. If an alligator is indeed T. rex adapted over time to the post-Flood conditions, then it would have similar DNA as a distant relative. There would be a slight difference because it's a copy of a copy of a copy, and so on. However, the identifying markers in the DNA depicting the kind of animal would be the same because the kind didn't change. Therefore, there is plenty of time for adaptation to modify T. rex to an alligator if they are the same kind.

The problem with evolution is not time. It doesn't matter how much time is allowed, you can use trillions of years, and it won't matter. Evolution utilizes random, unguided mutations of the DNA code for changes in function and kind. The odds of DNA code producing meaningful new functions that weren't already in the DNA code is beyond possible. Remember that the odds of random unguided changes to amino acids just for one single protein is 1 in 2×10^{195}. And evolution would need trillions of successful unguided changes to pull off some result in changes to different functions, let alone kinds. However, the DNA coding needed for adaptation is already embedded in the DNA code. With adaptation, the information is already in the DNA code in terms of how to handle certain changes in the environment. And when those changes are too much for that creature, guess what? It dies. It doesn't pass on any information to another creature. A common mistake by evolutionists is to see adaptation and confuse that with evolution. They are not the same, and this is not an exercise in semantics.

There doesn't need to be millions of years with adaptation because the information to modify a feature, or adapt in some way, is already in the DNA code, and creatures always stay the same kind though with many variations such that man may categorize them as different species. With adaptation, there is plenty of time from after the Flood to before Christ to have T-Rex adapt to become an alligator and for velociraptor to adapt and become a crocodile. The problem with the evolutionary hypothesis is a lack of information to begin with for the first living organism and a lack of information in the DNA code to change a function or kind of creature.

Ask evolutionists if they can name you one example of a mutation that increased genetic complexity. There is no naturally occurring example, evidence, or observation of a mutation that caused an increase in genetic complexity. Why? Mutations degrade the genetic code, not enhance it. It should be noted that <u>evolution doesn't require just one lucky break in the billions of years for the success of evolution. Quite the contrary, evolution requires trillions upon trillions of random unguided mutations to enhance the genetic code for the successful occurrence of evolution, yet there is not one example in nature.</u> This is a glaring fly in the ointment.

The concept of random, unguided changes to the DNA code is analogous to random, unguided changes to a software program, such as one that is designed to display a simple background color on a computer screen. Given enough time, would random changes in the code produce another functioning program, such as a word processor? The reality is that one single error in coding results in the entire program not functioning. And that is what happens to life; when there are errors in the DNA code, the life form doesn't function properly. There are stillbirths, sterility, premature deaths, physical mutations that prevent mating, weaknesses in defense mechanisms against predators, and so on. A mutation that has enhanced DNA coding has never been seen in nature. No one has observed mutations that led to a new function or that resulted in a new kind of creature.

If alligators are T. rex, they would be variations of the same kind, and though mankind would classify them as different species, they would still be the same kind (Family), with the same image and the same characteristics.

Review: If the alligators are a descendant of T. rex, they should have nearly similar DNA because they are copies of copies. This process would be through adaptation, which means the DNA necessary to adapt would already be in the DNA. The process of evolution doesn't have the information in the DNA to change and relies on the impossible means of random, unguided changes to DNA code to produce new functions and new kinds. *Adaption* and *evolution* are vastly different.

It's plausible that some dinosaurs are still living amongst us, but now they have been affected by reduced oxygen, increased solar radiation, decreased buoyancy force, and increased gravity and are not as large as their fossilized ancestors. This explains why many cultures have dinosaur drawings, engravings, and the like of dinosaurs before modern mankind had discovered such dinosaurs in archeological digs. While the dinosaurs were getting smaller and smaller, they were observed by ancient man and documented in carvings millennia after the Flood. For example, in the walls of the Angkor Wat Temple, an identical replica of a stegosaurus was carved in the stone walls. It seems a plausible explanation that they knew what stegosaurus looked like because they saw it firsthand.

Chapter Summary: Before the Flood, human beings lived 900+ years. After the Flood, the length of human life dropped dramatically with each generation. What changed? Oxygen levels went down, solar radiation was no longer shielded, the buoyancy force declined, and net gravity increased. This would also affect the dinosaurs. The Bible is clear that one of both male and female of every creature (including dinosaurs) entered the ark. Therefore, dinosaurs survived the Flood. And ~six generations later, in the Book of Job (chapters 40 and 41), two distinct dinosaurs are described in the present tense as alive and well: the behemoth and the leviathan. With the environmental changes, life span declined for all life forms, so dinosaurs didn't live as long either. Since dinosaurs are reptiles, we know reptiles continue to grow as long as they are alive. Therefore, if a life is cut short, then their size would dramatically be reduced as well. Their primary growth characteristics would be reduced and so too would their secondary characteristics. The saber-toothed tiger became the modern-day tiger. The mighty T. rex adapted and possibly became an alligator, and velociraptor adapted and possibly became a crocodile. Megalodon's size was reduced to the size of the great white shark. Mammoths slowly lost their hair and adapted to become the elephant. The size of the stegosaurus and its spinal plates were reduced to potentially a tuatara. The conclusion is that several dinosaurs are still living amongst us. They are not extinct; they just adapted to the new environment just like humans, who are also not extinct.

Group Discussion:

1. Seeing the similarities of certain dinosaurs with their descendants, do you think it is plausible that dinosaurs are living amongst us, albeit much smaller? And how does this affect your faith in the Word?

2. As God promises to restore creation back to a pre-fallen state, do you expect to see dinosaurs on Earth after sin and Satan are finally conquered?

Chapter 26
The Big Bang Versus the Bible

Does the Bible support the "Big Bang"? Yes and no. The Bible explicitly records that God stretched out and stretches out the heavens, which means the Bible implicitly declares that the universe was small at one point in time. Consider Isaiah 40:22: "He who sits above the circle of the earth . . . who <u>stretches out the heavens like a curtain, and spreads them out like a tent to dwell in.</u>" Also, consider Isaiah 45:12: "I have made the earth, and created man upon it; I, even my hands, <u>have stretched out the heavens and all their host have I commanded.</u>" There are about 17 verses that depict God spreading out or stretching out the heavens. It's clear in Scriptures that God spread out the heavens, which means that before He spread them out, they were closer in proximity. When was Isaiah written? Oh, about 700 BC, so this indicates that Isaiah supernaturally knew the earth was a circle (the word *sphere*, or *sphaira*, wasn't invented until the Empire of Greece) and that the heavens were expanding, some 2,000 years before science confirmed this information. From deep space, no matter the angle from which one views the earth, it appears as a circle. Only a spherical planet will appear as a circle from every angle in space. Thus, through deduction, the Bible was ahead of its time by demonstrating that the earth was not flat.

Some of the verses about God stretching out the universe are in the present perfect tense, explaining that what He did at creation is ongoing. For example, Isaiah 45:12 says, "I . . . have stretched out the heavens." This indicates a past event at creation in Gen. 1, on the second day of creation, that is also in the present tense, such as the universe is still expanding. Can God have it both ways—that He has stretched out the heavens and is stretching out the heavens? It turns out that when God stretched out the heavens, this set them in motion, and astronomers have discovered what the Bible proclaimed a 2,700 years ago—the universe is still expanding. Therefore, God's testimony is correct. How did the Bible know this 1,000 years before mankind?

Is this a violation of physics? Sir Isaac Newton discovered a law of physics called the first law of motion. It tells us that an object at rest will remain at rest, and an object in motion will continue in motion with a constant velocity unless it experiences an external force. How did Isaiah know to write that God stretch*ed* out the heavens and God stretch*es* out the heavens? Isaiah wrote this in 700 BC, and Sir Isaac Newton didn't pen the first law of motion till the early 1700s. Therefore, Isaiah wrote this 1,000 years before mankind new about the law. This is another illustration of the supernatural, divine nature of the Bible.

The Bible describes the conditions of the universe in the beginning before it was stretched out and before the sun, moon, and stars were finished coalescing in the universe on the fourth day of creation. And that condition allowed for liquid waters. Gen. 1:2: "The earth was formless and void, and darkness was over the surface of the deep, and the Spirit of God was moving over the surface of the waters." For water to exist in the universe before the formation of the sun and stars, the universe had to be small enough to exert enough pressure to generate heat to keep the waters liquid. But the universe couldn't be the size of a dot or what cosmologists call the singularity (the point at which all the matter in the universe was the size of a dot existing as energy), because at the point of being anywhere close to a singularity, all liquid would have long since vaporized, and so too would all other matter at that point, leaving nothing but energy.

The Bible describes liquid water existing before God stretched out the heavens and before the sun and stars were made, which means that the universe was small enough to exert pressure that generated heat to keep some of the waters in liquid form, but not too small as to vaporize all water. This is great evidence from a Biblical perspective that the Bible is against the singularity, but not against the Big Bang. What size would the universe have to be to have enough pressure to generate enough heat to keep the water liquid in space without any sun or stars? This is unknown; it could be the size of hundreds of galaxies side by side, or it may be the size of a thousand galaxies. The size of the small universe on the first day of creation, plays a critical role is determining the age of the universe. If the universe was the size of the singularity, then the Bible is wrong, and the evolutionary time line is correct. If the universe started from a much larger volume, then the Biblical time line is correct, and the universe is young.

It is interesting to note that scientists have just found out that the universe is expanding (in 1925, Edwin Hubble), which means that the universe was smaller in the past. And relatively speaking, mankind has recently found out that exerting pressure generates heat (late 1700s, Charles Law), which melts ice. So how did Moses, who wrote the Genesis account of creation 3,500 years ago, know that the universe was stretched out at creation—and that before the universe was stretched out and before the universe had the sun and stars—that water could exist in liquid form with a small universe exerting pressure to generate sufficient heat to maintain liquid water? I doubt that Moses knew any of that. Many times in Scriptures, the writers had no idea what they were writing about. Daniel had to ask several times to the angel what the angel was showing him. Daniel 12:8: "As for me, I heard but could not understand." And Daniel 8:15: "When I, Daniel, had seen the vision, I sought to understand it; behold, standing before me was one who looked like a man. And I heard the voice of a man . . . called out and said, 'Gabriel, give this man an understanding of the vision.'" Again, this demonstrates the supernatural divine nature of the Bible. The writers of the Bible wrote scientifically accurate information that is in harmony with physics some 3,500 years before mankind had any idea that there was a relationship between volume, pressure, and temperature. This leads the reader of the Bible to the real author, God. "All Scripture is inspired by God" (2 Timothy 3:16).

Review: The Bible implicitly declares that the universe was small before God stretched it out. The Scriptures record that the waters existed before the creation of the sun and stars, establishing that the universe was small enough to generate pressure and heat to maintain water in a liquid state. But there was not too much heat as to vaporize all the water. Hence, the Bible is at odds with the singularity and old universe hypothesis, but not with the Big Bang.

When did God stretch out the universe? The universe (the second heaven in relation to earth) was most likely stretched out on the second day of creation and resulted in the atmosphere being stretched out. Note that the term *heavens* was pluralized in Gen. 1:9 (on the third day of creation). On the fourth day of creation, the sun, moon, and stars are commanded by "let there be lights in the expanse of the heavens." The expanse of the heavens/universe seems to already be there for the sun, moon, and stars. It is clear though that on the fourth day the sun, and stars were ignited, and the planets and moons were finished coalescing. Gen. 1:14:

> Then God said, "Let there be light in the expanse of the heavens to separate the day from the night, and let them be for signs and for seasons and for days and years; and let them be for lights in the expanse of the heavens to give light on the earth"; and it was so. God made the two great lights, the greater light to govern the day, and the lesser light to govern the night; He made the stars also. God placed them in the expanse of the heavens to give light on the earth . . . There was evening and there was morning, a fourth day.

Review: The Bible explicitly declares that God stretched out the heavens, which occurred on the second day, indicating that before God stretched out the universe, the universe was small.

The Bible does support a very small universe at one time, but this is in direct conflict logically with the premise of a singularity and old universe hypotheses. What's the difference between the Bible's "Let there be an expanse" and the evolutionary "Big Bang"? Scientists looked at the universe and determined that the universe was still expanding. This means that the further back in time we go, the smaller the universe becomes, and if one continues further back in time beyond the time limits indicated in the Bible, one arrives at a point in time where the universe cannot get any smaller. This is how one gets to the notion that all matter (existing as energy) in the universe was the size of a dot in the beginning of time. There is nothing in physics, cosmology, astronomy, the Bible, science, chemistry, or anything that says one has to go back in time until all the universe was the size of a dot unless one believes there is no God, or one believes the Bible to contain errors. Then, one has to go back in time to the point of a singularity to make sense of the natural processes that are observed today.

It makes sense to go backwards in time till the universe was smaller, but not to go back too far in time so that it would require too much energy to reduce the universe smaller and smaller to the size of a dot. There is a point in time, when going back further in time, at which it requires too much energy to further reduce the size of the universe. No one knows the area/volume of the small universe before God expanded it because the rate and duration of universal inflation is not knowable. Thus, by faith, creationists accept the idea that God chose to stretch out His heavens on the second day, and He did it according to His word. And by faith, evolutionists accept the idea that all the universe existed as a singularity and the universal inflation after the Big Bang was one billionth of a billionth of a second.

How do astronomers determine that the universe is expanding? Here's a little bit of fun on how scientists first discovered that the universe is expanding. The year is 1925, and Edwin Hubble, utilizing the Doppler effect, determines that the universe is expanding. What is the Doppler effect? Austrian physicist Christian Doppler, in 1842, determined that there is a change in frequency for the wavelength of an object heading toward an observer versus the frequency for the wavelength of an object traveling away from an observer. For example, take a duck swimming in a pond. The waves of the water (wavelengths) in front of the duck are compressed (higher frequency and smaller distance of wavelength), and the waves (wavelength) behind the swimming duck are elongated (lower frequency and larger distance of wavelength). Therefore, just by looking at the waves the duck makes, one can determine which direction the duck is swimming.

Similarly, the sound of a siren on a police car traveling toward an observer has a higher pitch (sound waves are closer together, with higher frequency and higher pitch) versus the sound of a siren on a police car that is traveling away from an observer, which has a lower pitch of sound (sound waves are further apart, with slower in frequency).

Similarly, light waves work on the same principle of physics. Light waves from a source traveling toward an observer are higher in frequency (blueish), and light waves from a source traveling away from an observer are slower in frequency (reddish). Utilizing Austrian physicist Christian Doppler's work, Edwin Hubble observed red and blue light emanating from different sources. Applying the Doppler effect, Hubble determined that since the source of a distant star is red shift, then the source of the light is traveling away from the observer, leading to the idea that the universe is expanding.

Review: The universe is expanding, which means the universe was once smaller.

How do cosmologists get to the notion that the Big Bang occurred ~14.6 billion years ago? They do it by estimating the relative distance traveled so far and dividing it by the estimated rate of expansion of the universe today to get an estimated origin date. The Bible is in partial agreement with this. The Bible declares that on the fourth day of creation, God finished making the sun, moon, and stars, and placed them in a stretched-out heaven, in this case, the universe. Therefore, by working backward in time, one can surmise that there was a smaller universe in the past using both the Bible and the Big Bang hypothesis. However, creationists contend the beginning was around 6,000 to 10,000 years ago, and evolutionists contend the beginning happened around 14 billion years ago. The problem is that no one knows the rate or the duration of the universal inflation that occurred from the Big Bang. And no one knows the initial size of the universe in the beginning before the universal inflation. As we discussed earlier, evolutionists contend that there was a time, called universal inflation, when the universe expanded faster than the speed of light. Evolutionists contend that the duration of universal inflation, for our universe, was a fraction of one second. But their estimate of the duration is a guess, and if the duration of the universal inflation, lasted a day, and not one billionth of a billionth of a second, then the universe would be very young, potentially 6,000 years old. Therefore, an evolutionist's hypothesis of the duration of the universal inflation is no more valid than God's testimony of how long He accelerated the expansion of the universe. Both evolutionists and creationists have the same precept of a Big Bang, though they use different nomenclature to describe a similar event. The "expanse" on the second day of creation, has a violent root definition, like a blacksmith hammering out iron. The imagery is of sparks flying, glowing hot metal, steam, violent force, and so forth. Thus, we may conclude that the Big Bang is a close representation of the expanse on the second day of creation. Amazing how the Holy Spirit guided Moses to choose that special Hebrew word that is associated with violent force, sparks flying, and rapid expansion. Creationists give credit to God for the violent expanse of stretching out the heavens, and evolutionists give credit to the seemingly natural processes of the Big Bang that stretched out the universe. A creationist will rightly argue that the design for the natural processes came from God, and this is where a theistic evolutionist smiles in agreement. Both creationists and evolutionists accept a moment of time of accelerated universal inflation immediately following the expanse or Big Bang. But they differ on the duration of the universal inflation and the starting size of the universe. There is no conflict with science regarding a young universe. There is only a conflict of a young universe concept with a man-made hypothesis called evolution and the ~14.6 billion years required for evolution. For if the universe is young, then evolution doesn't have enough time according to their slow evolving processes. The crux of the matter is this: Does one have faith in the singularity starting point for the Big Bang hypothesis, which suggests that the universe was the size of a dot with a brief—a billionth of a nanosecond, which is 10^{-18} second—universal inflation, or does one have faith in God's Word with a small universe the size of ~100–1,000 galaxies, a violent expanse (Big Bang) on the second day of creation, and a prolonged—a day—universal inflation? Cosmologists provide their best estimates, in the form of a hypothesis, but they weren't there to observe the duration of expansion in the beginning, and they can't test the rate or duration of the expansion in the beginning, and neither can creationists. But most importantly, cosmologists don't know the starting point or the size of the universe when it started expanding. From what starting point did the universe start expanding? This is where evolutionary scientists are no longer practicing science, but a faith-based belief system. Both are faith based as to the initial size of the universe, the date of the beginning, the rate of universal inflation, and the duration of the universal inflation. Creationists admit to having faith, and evolutionists try to say they don't have faith because it is fact. Both views cannot be correct, and the authenticity of both faiths hangs in the balance.

At the beginning of time when the universe first started expanding, if that starting point was not the size of a dot as Big Bang theorists contend—but let's say the size of ±1,000 galaxies—and if the duration of the universal inflation was a second, then the universe is very young. If at the beginning of time when the universe first started expanding, if that starting point was not the size of a dot as Big Bang theorists contend—but let's say the size of 20–50 galaxies—and if the universal inflation was a day, then the universe would be young as the Bible implies. If at the beginning of time when the universe first stated expanding, if that starting point was the size of a dot, and the duration of the universal inflation was one billionth of a billionth of a second, then the universe would be around 14–20 billion years old and evolutionary cosmologists would be correct.

A key to unraveling this mystery are the waters in Gen. 1:2. For waters (solid, liquid, gas, plasma, and abundant) to exist in the universe before God stretched out all the matter and before He formed that matter into the sun, moon, and stars, the universe would have had to be small enough to generate heat and pressure to keep water in a liquid state, but not too small as to generate too much heat and obliterate all water. This rules out the singularity of the dot from the Big Bang hypothesis and leaves us with a universe at the beginning that is compressed, but not too compressed, potentially the size of 100–1,000 galaxies combined. This is a litmus test for theistic evolutionists—do you reject the Bible or do you accept the Bible? Waters existing before the Big Bang, before the expanse of the second day of creation, means the singularity hypothesis is wrong and the universe is young. For the foundation of an old universe is that the beginning of time started with a singularity. But since the singularity is impossible, according to the Bible, then the only viable solution is that the size of the small universe was vastly larger than evolutionists are purporting before the Big Bang. Otherwise, waters would not exist. And this takes away billions of years from their time line that is the foundation for evolution. Therefore, the Bible, with mentioning that waters existed before God expanded the universe in verse 2, single-handedly debunks an old universe construct, and by default removes the foundation for evolution.

The initial volume of the universe has to be within a range for water to exist in four forms—too small and then all the water is vaporized, and too large and then all water freezes to ice. The size has to be within a range for water to have enough pressure and heat to exist in the plural form (ice, liquid, gas, and plasma), with a gas form near coalescing, burgeoning stars, with plasma formed within the coalescing celestial bodies, with areas further away from the coalescing stars existing in liquid form, and with the outside portion of the universe existing in solid (ice) form as a shell. All four states of water played a key role in expanding the universe, or Big Bangs. And this is potentially why the Bibles pluralized waters in Gen. 1:2.

With no heat source on the outer rim of the small universe, the water located there would be ice. The ice acted as a thick outer shell to the universe, forcing pressure to build internally. Without this outer shell of thick ice, potentially thousands of miles thick, then no pressure would build up, and then there would be no Big Bang on the second day of creation, and no rapid release of pressure of the water that surrounded the early earth, and thus no atmosphere would have been formed on the second day of creation either.

The water nearest to the burgeoning, spinning, hot, coalescing stars, vaporized to gas. This process applied pressure inside the tightly packed universe for the Big Bang on the second day of creation. Without this water nearest the forming stars, then there would be no gas, and no pressure, and no Big Bang. Therefore, without water, the universe would still be tightly packed, and all the trillions of individual coalescing stars would combine into one molten mass, with no individual stars.

The water in between all the spinning masses provided the supply for either converting liquid to gas to increase the pressure nearest a heat source and supplied the conversion of liquid to ice to seal in the pressure furthest away from the heat. In this liquid medium of water, filling the tightly packed universe at the location of a formless and spinning earth that was void of life, we read of the Holy Spirit hovering over the deep (Gen. 1:2). The four states of water were made possible by trillions of spinning, coalescing celestial bodies at the beginning of the first day of creation, though darkness was everywhere. With each passing minute, trillions of coalescing celestial bodies gained mass and heat, until enough matter coalesced and enough heat was generated, that the future stars started to glow hot at the middle of the first day with, "Let there be light" (Gen. 1:3). The process of converting liquid water into gas by trillions of coalescing masses pressurized the small universe, and this pressure was sealed in by an external shell of thick ice. And this was a key component of the Big Bangs (on the second day of creation). With liquid metal coalescing, such as alkali metals that are highly explosive to water, this created a recipe for an explosive second day of creation.

With the trillions of glowing hot coalescing masses spinning, they caused a flow of electrons, and this process resulted in a buildup of unstable amounts of positive and negative charges on a cosmic level. This process is similar to an alternator, generator, or thunderstorm that builds up electric charges.

When the pressure was near an unsustainable load and the electric charges were at an unstable level, then God said, "Let there be an expanse in the midst of the waters," and lightning sprang forth throughout the universe; this was the final component, and this action on the second day of creation caused the Big Bangs. This rapid release of pressure from the Big Bang, caused our atmosphere to form on the second day of creation by a process called cavitation. This process can be observed by opening a new soda bottle. The spinning, coalescing masses (the future stars) continued adding mass until the fourth day, when they coalesced enough mass to ignite after God said, "Let there be light in the expanse of the heavens." Thus, God supernaturally used natural processes to fulfill His Word.

Review: When God created all matter for the universe on the first day of creation, when it was formless and void of life, and the Holy Spirit moved over the surfaces of the waters, the size of the universe at that moment provides a great clue to how old the universe is. If the universe was infinitely small like a dot on a page, then billions of more years are required to stretch out the universe to its current size. However, if the universe was the volume of 100–1000 galaxies, then 6,000 years are sufficient. No one knows the duration of universal inflation of the early universe, if it lasted only a brief Planck time, then billions of years is required. However, if the universal inflation lasted hours, then 6,000 years are sufficient. The answer is faith based because no one was there to observe. Gen. 1:2 records that waters and dirt existed, which means the universe was large enough on the first day for waters to exist in solid, liquid, gas, and plasma form, potentially the size of 100–1,000 galaxies. The bottom line is who do you believe? Man, who has been observing the cosmos for centuries, or God who created the cosmos. The mention of waters in the opening verses, rules out the singularity as the starting point, and solves the mystery of the age of the universe, that it is young.

By the way, it should be pointed out that no one knows how fast the universe is expanding. People can only surmise that the universe is expanding from viewing the galaxies separating. But no one knows what the ratio of universal expansion (outer edges) is compared to galactic expansion (inner material). It probably isn't a one-to-one ratio; it is more likely a sliding scale of faster expansion the closer to the edge of the universe. It's an unknown because no one can see the edge of the universe. But we might have a smaller scale model that may help us. When a grenade explodes, the highest velocity of matter are the initial particles traveling away from the central location, with a sliding scale of the slowest particles the last to leave the central location. Thus, the closer we visualize the outer edges of our universe, the higher the velocity of galactic expansion should be.

It is interesting to note that the creation record is told from the perspective of Earth. Therefore, it is plausible that our galaxy, the Milky Way, could be near the center of the universe. Cosmologists have determined that every galaxy is expanding from every other galaxy. Even though this is accepted, there still has to be a center of all the expansion. And that center could vary well be our solar system in our galaxy.

Does the Big Bang theory conflict with the Bible? Not exactly. The Big Bang theory suggests that all matter (as energy) was compressed to the size of a dot and exploded some 14.6 billion years ago. This explosion allegedly set in motion all the natural processes we see today, meaning that everything we see today is explained by natural processes, and there is no need for God. The Bible is in direct opposition to the singularity (the size of a dot) premise for the Big Bang by declaring that water and matter existed before the Big Bang of the second day of creation. And the Bible is in direct opposition to the evolutionary time line by saying that God created everything in six literal rotations of the earth (day) and rested the seventh day, with the genealogies suggesting that God's creation occurred some 6,000 to 10,000 years ago. And the Bible is in direct conflict with the order of the Big Bang. The Bible has matter being created before the expansion/Big Bang, and we may infer that the conditions of the matter caused the Big Bang/expansion. However, evolutionists believe that the Big Bang proceeded and formed the matter. Otherwise, the idea of a Big Bang with universal inflation is acceptable for creationists, as they parallel the second day of creation.

Can believers in God believe that God created everything through the Big Bang some 14.6 billion years ago and that the Bible is not exactly accurate, but merely exists to give us general ideas of what happened? They can believe this, but then they are calling God's Word a lie because God in the flesh declared that not one jot or tittle would pass from the law. God's Word gave details of His creation in Gen. 1 and 2 and repeated it (twice verbally and twice written in stone) with a general summary of how long it took God to create everything in Exodus 20 and 31—the law. And who does the Bible say is the Word of God? That would be Jesus, the co-creator, the second member of the one God, the Word of God became flesh and dwelt amongst us (John 1:1–18). Therefore, those who claim to believe in God, but declare His Word to not be accurate, are going against Jesus. Jesus said you are either for Me or against Me (Matthew 12:30), so those who claim that the beginning of time happened according to how evolutionary cosmologists say how and when it happened and not how God testified how and when He spoke them into existence are calling God a liar.

God testified that His Word is perfect, sure, right, pure, clean, and true (Psalm 19), so when God says He made everything in the heavens and on earth in the six days of creation, it is a litmus test of ones faith. But take heart because each event in the Genesis creation account is explainable with science. For example, utilizing how scientists at the C.E.R.N. Hadron Collider converted energy into matter, is a plausible explanation for how God created matter on the first day of creation. And the principle in physics called cavitation to explain how God made the atmosphere from the pressurized water, that once surrounded the earth, on the second day of creation. And utilizing geologist's explanation of how land is rising today with Post-Glacial rebound, to explain how God gathered the waters together and made the dry land appear on the third day of creation. And the light at the middle the first day of creation with molten, liquid matter as the sun was coalescing matter, until the fourth day when enough matter had coalesced that the sun was able to fuse hydrogen.

Therefore, a believer in God can accept the expansion of the Big Bang, but not the concept of the singularity, or that matter didn't exist prior to the expansion/Big Bang (second day of creation), or the old earth/universe hypothesis. In fact, scientists have discovered signatures in the universe as cosmic background radiation that suggests past explosions. The prevailing hypothesis is one Big Bang, but viewing the ubiquitous nature of the cosmic background radiation indicates that the one Big Bang was comprised of trillions of simultaneous smaller Big Bangs.

Review: It all comes down to the size of the compressed universe before God started expanding it. If the universe was the size of 1,000 galaxies combined, then the duration of the accelerated rate of universal inflation was shorter. If the universe was the size of 20–50 galaxies, then the duration of the accelerated rate of universal inflation was longer. Either size is acceptable, as long as the compression of a small universe generates enough heat to keep water in a liquid state, but not too compressed to vaporize all matter. Both rule out the singularity of the Big Bang hypothesis.

Group Discussion:

1. Knowing that the principle of the singularity is exclusively by faith alone, and is incompatible with Scripture, how does this shape your understanding of the origin of all things?

2. Since the Bible records matter and water were created before the expansion/Big Bang, and since it is by faith alone that evolutionary cosmologists believe the Big Bang formed all matter, then where should you put your faith, in the Word or in man's philosophies?

Chapter 27
The Big Bang Versus Physics

Does the Big Bang conflict with physics? The second law of thermodynamics is called entropy. And it says that everything goes from order to disorder, from complex to simple, and heat flows toward a cooler object.

Big Bang theorists believe that the Big Bang exploded from the spinning hot massless energy of the universe, which was the size of a dot, and from that, all the natural processes we see today evolved. Everything—even the complexities of the DNA code, the information to form life, consciousness, and morality—came from chaotic disorder. To have all the universe squished to the size of a dot is considered orderly. And exploding the universe could be considered massive disorder. So far so good, right? But we have skipped a step. How and where did all the energy come from in the singularity? What is the intelligent source required to squish the universe to an orderly dot? This formation of all the energy and matter of the universe squished to an orderly dot is a violation of entropy. Before the Big Bang, everything went from disorder to an orderly dot of pure energy, spinning at high velocity called the singularity. That is backward to the law of entropy, which states that everything goes from order to disorder. The Big Bang hypothesis that everything in the orderly singularity explodes and goes from the disorderly explosion at the Big Bang to the orderly formation of all the galaxies is another violation of entropy. Going from disorder to order is backward and violates entropy.

Now, there are instances of going from disorder to order in a smaller system, when an outside source (we'll call this outside force "A") applies a force, which must include intelligence to utilize the force to that disorderly system (we'll call the disorderly system "B"), but the overall entropy of both systems combined tends toward disorder. If we applied fictitious numbers to this principle of physics, it would look like this:

A: The outside source applying order is "A." Value: 100 equals highly ordered.
B: The disorderly system being modified is "B." Value: 10 equals disorderly.
C: The combined overall value is 110.
D: The cost of applying order is a two-to-one ratio.

Where "A" (the outside source with a value of 100 units of force) applies 20 units of force to "B" (disorderly system with a value of 10), the resultant change to "B" is +20 units of force, resulting in "B" having a value of 30. But the cost to "A" applying the force equals –40. Thus, the changes are:
"A" outside source: Value of 100 reduced by –40 = net 60, a decrease in order.
"B" disorderly system: Value of 10 changed by +20 = net 30, an increase in order.
C: The combined overall value is 90. Entropy increased as the two systems combined has less order.

Therefore, although "B" became more orderly, the overall combination of both systems tended toward disorder, which is in accordance to entropy. Creationists need to stop saying that order can not arise from disorder because if an outside system applies energy, then it is doable. What creationists should say is that even though order may increase when a force is applied, the overall order will always tend toward disorder, and always intelligence is required. Entropy always increases.

To most honest evolutionary cosmologists and theoretical physicists, the enigma of the Big Bang is an unknown. The most honest explanation for the belief in the Big Bang comes from theoretical physicist Michio Kaku, who said, "We don't know <u>why</u> it banged, we don't know <u>what</u> banged, and we don't know <u>how</u> it's banging." Honest evolutionists admit they don't know the "why, what and how" of the Big Bang; however, most evolutionists don't speak in such honest terms—instead, they proclaim evolution and the parameters of the Big Bang as fact. Take for example

Lawrence Krauss, a cosmologist who explains the Big Bang not as a hypothesis, but as fact—as "the moment when everything we see, all hundred billion galaxies, each of which contains a hundred billions stars, all that material, was compressed in a region that was infinitely small." This is a summary of the singularity. And it is in complete opposition to the Bible that records waters and dirt were created before the God stretched-out the heavens with a Big Bang. This is one reason a believer in the Word—Jesus (John 1:1–18)—cannot be a theistic evolutionist. They must choose either to believe the likes of Lawrence Krauss, or believe the Word of God.

Where and what is the orderly force that applied order to the disorderly universe that just exploded from the Big Bang? Some speculate the existence of multiple universes applied the necessary force to our universe. Some attempt to solve the problem by adding a large enough time coefficient. Often, I have read suggestions from evolutionists, such as given enough time, what is impossible becomes possible, and the possible becomes probable, and the probable becomes certainty, and the certainty becomes fact. What evolutionists are saying is that they concede it's impossible, but given enough time, the impossible is not only probable, but it becomes fact. It never ceases to amaze me the faith that evolutionists have. The two belief systems are similar, as the Bible says, "With man this is impossible, but with God all things are possible." If this wasn't such an important topic, it would be comical how similar they are at times; both require faith, and both are a belief system. Christians admit they have faith and a belief in God, and atheists pretend they have no faith and no belief system—just unbiased science and empirical data and facts. The funny thing is that believers in God see that atheists utilize extreme amounts of faith and have a very strong belief system, but atheist can't or won't see it.

The Bible foretold of mankind pursuing myths of creation and rejecting the Bible's Genesis creation account (Romans 1:18–23) and chasing after their own desires and positioning teachers to teach them what they want to hear to tickle their own desires (2 Timothy 4:3–4). The Bible describes that God allows them to pursue their own desires and hands them over to a depraved, non-functioning mind (Romans 1:24–28).

The Bible does not violate any laws of physics, including entropy, in regard to any aspect of Genesis creation, but specifically as it relates to the origin of the universe. In fact, entropy is supported in the Scriptures. It appears almost a millennium before science caught up to speed. Isaiah 51:6 and Hebrews 1:11 both explain that Heaven and Earth will wear out like an old garment, and 2 Corinthians 4:16 explains the outer man is decaying, which is entropy in a nutshell. This is further evidence that the Bible was written by the inspiration of God, by talking about sound physics and knowledge of entropy 1,000 years before mankind had a clue.

Does the Bible's creation account violate entropy by going from the disorder of all the water and dirt being formless and void of life in Gen. 1:2 to God creating extreme amounts of order through six days of creation? No. God is the outside source applying force to the disorderly system. Then, does God reduce in order? No. God is infinite and all powerful, so there is no reduction in infinity (God's order). If we applied fictitious numbers to this scenario, it would look like this:

A: The outside source applying order is God. Value: infinite.
B: The disorderly system being modified is creation. Value: 0.
C: The cost of applying force is a two-to-one ratio of cost versus change. The overall value of the two systems combined is infinite.

Infinite God (the outside source) applies 30 units of force to B (the disorderly system, creation) with a value of 0). The resultant change to B is +30 units of force, resulting in B being at a value of 30. The cost to A applying the force equals –60. Therefore, the changes are:
A: God: infinite value reduced by –60 = net infinite.
B: Creation: value of 0 changed by +30 = net 30.

C: The overall value of the two systems combined is infinite.

It's a simple mathematical model, but when something is infinite, there are no changes to its value, strength, or properties. That's why the Bible is consistent with physics when the Bible declares that God is the same yesterday, today, and forever (Hebrews 13:8)—that God is infinite. All that God created is subject to entropy—the law is not, the truth is not, and the Word of God is not.

I was asked by an atheist, "Do we have the same DNA as God?" Since all of creation was made from God, everything would have some signature of God, some more than others, with humans having the closest DNA autograph. Since we are copies of copies with respect to DNA, then the further back in time, the closer our DNA gets to God's, with Adam and Eve having the closest replica of God's DNA. This is also in compliance with entropy.

Review: *Entropy* **means that all things tend toward disorder. The singularity of the Big Bang hypothesis violates entropy by the idea that the universe organized itself into an orderly singularity (a spinning dot). It explodes into disorder (which is acceptable) and then proceeds to the order of all the galaxies we see today; that's another violation of entropy. But the Bible is in harmony with entropy and discusses the concept of entropy almost 2,000 years before mankind found out about this law of physics.**

Another aspect of entropy that the Big Band theory violates is the design or intelligence involved in the application of force to move from disorder to order. Entropy states that all things tend from orderly to disorderly. The only way to go from disorder to order is to have an outside source apply a design force or intelligence involved in utilizing the force.

Why does design or intelligence have to be involved in the process of adding more order? Have you ever tried to clean your room by throwing a stick of dynamite in the room? Nothing organized can come out of an unintelligent force without design. But that is exactly what Big Bang theorists believe. The Big Bang, which is an unintelligent, non-designed force, fostered all the natural processes of nature that are intricately fine-tuned—even the laws of physics—to form the galaxies and eventually all of life as we know it. This is a violation of entropy; something can't go from disorder to order without an outside source and intelligence utilizing the force applied. Everything tends toward disorder. How do evolutionary Big Bang theorists get around this? They cite the multiverse. They believe that an outside source from our universe is the fuel source for our universe to go from the disorder of the Big Bang to the order of the laws of physics, chemistry, cosmology, math, and biology and the complexities of life.

It should be noted that a multiverse is pure speculation. The evolutionists will mock creationists for their alleged blind doctrinal faith and ridicule them for allegedly going against science. The reality is that creationists and the Bible never go against science, but they frequently go against evolutionary scientists regarding creation and the origin of life.

It is sad is to hear of believers in God proclaiming that they believe in man's philosophies versus the Word of God. Yet, the Word of God is the one in harmony with physics by declaring that all of God's creation obeys entropy (Isaiah 50:6 and Hebrews 1:10–11). "You, LORD, in the beginning laid the foundation of the earth, and the heavens are the works of your hands; They will perish, but you remain; and they all will become old like a garment . . . But you are the same, and your years will not come to an end." The problem is the heart of man, who utilizes his own intellect to interpret data instead of the mind of Christ (I Corinthians 2:16). That's the problem; the problem has never been science. Science is a dear old friend to a creationist in the search for truth. Since science means to know, and physics is for us to understand all the workings in the universe, then Proverbs hits the nail on the head with, "The fear/reverence of the LORD is the beginning of knowledge (science)" (Proverbs 1:7). And Proverbs 9:10 says, "The fear/reverence of the LORD is the beginning of wisdom, And the knowledge of the Holy One is understanding." Therefore, the first steps to know and understand all the workings of God's creation, one must fear/reverence the LORD. Even a child that accepts the Genesis creation account can speak profound knowledge with understanding that is beyond the atheistic, erudite elites—just from simply reverencing God.

Review: The Big Bang theory violates physics by going from the disorder of the Big Bang to the orderly fine-tuned universe, utilizing no intelligence in the application of an outside source to evoke the necessary order we see in the fine-tuned universe.

There is another law in physics that is very important to discuss that concerns the hypothesis of the Big Bang. It's the first law of thermodynamics, and it deals with the conservation of energy. Energy comes in many forms, such as light, heat, chemical, electrical forms, and matter. The first law of thermodynamics says that energy can change forms, but it cannot be created or destroyed, and therefore, the total sum of energy and matter remains constant throughout the universe.

How is this a problem for the Big Bang? Where did the energy come from to compress the universe to the size of a dot, fill the singularity with infinity energy, and to spin the singularity for the Big Bang?

This is a problem for evolutionary Big Bang theorists because they either have to believe by faith that all matter and energy in the universe spontaneously evolved at some point in time or all matter and energy existed eternally.

Some evolutionists believe that nothing existed before the Big Bang, and then the Big Bang and everything evolved out of nothing. This hypothesis has an increase in the total sum of energy. This is a violation of the first law of thermodynamics and is a belief construct that is not science.

Some evolutionists believe that all matter and energy existed before the Big Bang; if so, then who/what created them? If matter and energy eternally existed, then this requires faith like a religion and is not science. Let that sink in for a bit. We know it's hard enough to squish a sheet of paper to the size of a dot, let alone all matter in the universe. The amount of energy and design/intelligence required to compress all the matter in the universe is beyond calculations, and the heat generated from such a compression would consume the matter that was compressed. In other words, the matter would be used up to generate the heat because heat can't be spontaneously created without something being used up. In this case, at the singularity, the matter would have been used up to generate the massive heat. The heat generated would be in the order of around $\sim 1.0 \times 10^{31} °C$, which would incinerate the matter.

Some evolutionists believe that only energy existed at the time of the singularity, and as a result of the Big Bang, then massless particles colliding at the speed of light, formed all the mass of the universe we see today. Where did the energy come from? and what caused the orderly design of the singularity? This violates the first law of thermodynamics with the initial creation of the intelligent energy that formed the singularity and the creation of the energy contained in the singularity. And if evolutionists argue that the energy eternally existed, then this is faith based like a religion.

Remember, nothing comes from nothing. The evolutionary response is that just because they do not know where the energy came from, doesn't mean the theory is wrong, nor does it mean that creation is correct. Evolutionists solve the origin of the energy with a belief that multiple universes provided the energy. A multiverse is a leap of faith, the existence of only energy as a singularity is a leap of faith, and believing all matter formed from the Big Bang's energy is a leap of faith. This is fine to believe, but don't call it fact or science; a multiverse is not observable or testable. And we are back to the same problem, where did the energy come from to form the singularities and fill those singularities with energy in all the other universes. Eventually, evolutionists are left with a faith-based hypothesis or violate the first law of thermodynamics by believing that there was an increase in the total sum of energy and matter at one point.

So the problem with the Big Bang hypothesis is that it violates the first law of thermodynamics. Big Bang theorists want people to believe that all the energy to squish the universe to a dot, and heat the dot, and spin the dot, and explode the dot, not only survived, and didn't produce inert ash, but formed all the matter in the universe with ample energy left over. This energy somehow spontaneously evolved or came from an outside source, such as other universes. This is where science morphs into a faith-based endeavor and departs from the observability requirements of the scientific method.

Review: The Big Bang violates the first law of thermodynamics, which states that the sum of energy and matter remain constant. Therefore, the increase in the total sum of energy to infinity is a violation of the first law of thermodynamics. If Big Bang theorists believe that the energy eternally existed, then this is faith based and is not science. If evolutionists proclaim that nothing existed in the singularity, then the increase in the sum of matter and energy from the Big Bang is also a violation of the first law of thermodynamics. Either they violate the law of physics, or they reveal their hypothesis is faith based.

How many times in your life have you had to try and try to get something correct? Rarely does something work out perfectly on the first attempt. Take a beginner at any activity, how many times will it take them to achieve perfection? Using the scientific method, we observe that in almost everything we do, there requires some practice, some repetition, some redoes, and some failures to get it right, but not with the Big Bang. Oh no, on the first try, the Big Bang perfectly laid the foundation and set in motion all the laws, constants, and natural processes that eventually led to spontaneous life. Boy, we sure are fortunate to have gotten lucky with that accidental perfection. Imagine what kind of pickle we would be in if the Big Bang was too little of a bang or too big of an explosion.

This perfection on first attempt doesn't seem realistic. Even God had a redo with destroying mankind with the Flood and sending His Son to die on a cross for the sins of mankind. That seems to be the case, but it was mankind that needed the redo, not God. God actually planned out prior to creating anything that His Son would die for mankind to redeem mankind from their sins. God, with foreknowledge, knew that mankind, with free will, would fall short of His glory and be deserving of the penalty of sin, which is death. This demonstrates the mercy of God to withhold instant judgment on sinners who are deserving of instant death; this demonstrates the justice of God in rendering righteous judgments. This demonstrates the love of God in sending His unique, one-of-a-kind begotten Son to bear the full cost of sin for those who love Him. This demonstrates the grace of God to give freely, full forgiveness of the cost of sin to those who ask of Him. Where is it that God predetermined to crucify His Son before time began? Acts 2:22–23: "Jesus the Nazarene. . . delivered over by the predetermined plan and foreknowledge of God, you nailed to a cross." God doesn't increase in knowledge because He knows all things, so the predetermined plan was in place "In the beginning."

The universe consists of finely tuned laws. Slightly altering one aspect of the fine-tuned laws would collapse the other finely tuned laws, which would eventually collapse the entire universe as we know it. For example, consider a slight alteration in an electric charge of atoms. If the charge is slightly weaker, then the nucleus of atoms would not be stable to hold onto electrons. If it's slightly stronger, then the electrons would collapse on the central nucleus. Either scenario would have prevented life as we know it. Nature's physical constants are so precise that it is improbable that they were set in motion by the Big Bang. If gravity were weaker so that matter couldn't coalesce to form planets or stars, or if it were too strong so that the attraction formed only black holes, then life couldn't be sustained or initiated. If gravity was too strong at the Big Bang, then the universe would have collapsed back on itself in a "Big Crunch." If gravity were too weak, then all the galaxies and stars would have continue to expand too fast from each other, and we would have had a "Big Freeze." Either would have meant that life couldn't exist. Evolutionary scientists place all the myriads of fine-tuned laws within our universe in a "Goldie Locks" scenario. Every parameter, every law, every constant, every element, and everything needed for life on earth was "just right." Take for example the mathematical form that nature takes on. Albert Einstein had this to say after observing the mathematical precision and harmony of the universe and all that is within it:

> I consider the comprehensibility of the world (to the degree that we may speak of such comprehensibility) as a miracle or an eternal mystery. Well, a priori [deductive reasoning from self-evident propositions], one should expect a chaotic world, which cannot be in any way grasped through thought . . . The kind of order created, for example, by Newton's theory of gravity is of quite a different kind. Even if the axioms of the theory are posited by a human being, the success of such an enterprise presupposes an order in the objective world of a high degree, which one has no a priori right to expect. This is the miracle which grows increasingly persuasive with the increasing development of knowledge.

Dr. Walter Bradley, who has a PhD in material science and a BS in engineering science, while discussing the strong nuclear force of atoms, had this to say:

> If the strong force which binds together the nucleus of atoms were just five percent weaker, only hydrogen would be stable and we would have a universe with a periodic chart of one element, which is a universe incapable of providing the necessary molecular complexity to provide minimal life functions. . . On the other hand, if the strong force were just two percent stronger, very massive nuclei would form, which are unsuitable for the chemistry of living systems.

And the amount of matter also reveals how finely tuned the universe is. If there was too much mass, then the universe begins to collapse upon itself because of gravity. Or if there was not enough matter, then the universe would have expanded too fast and froze. A critical density is required for equilibrium to prevent the universe from freezing to death or collapsing upon itself, and to paraphrase Michio Kaku, theoretical physicist, "That critical density is ~5 hydrogen atoms per cubic meter average throughout the universe. If there were more hydrogen atoms, then the universe would begin to collapse; if we had fewer hydrogen atoms, then the universe would expand to super freeze death."

The finely tuned universe cries out that only an all-powerful God, knowing exactly the measurements necessary and requirements for life to be sustained, could have created everything. Psalm 19:1–3: "The heavens are telling of the glory of God; And their expanse is declaring the work of His hands. Day to day pours forth speech, and night to night reveals knowledge. There is no speech, nor are their words their voice is not heard." The universe testifies for the almighty Creator.

Scientists have discovered that there is dark matter filling what was previously thought of as empty space between galaxies. Scientists studied 42 supernovas over an eight-year period and discovered something remarkable. When the stars exploded, the explosion continued accelerating as though something was propelling it faster and faster. We don't know what this matter is, but scientists call it "dark energy/matter." Some have suggested that dark matter has a repulsive gravity—a kind of anti-gravity that pushes all galaxies away from each other, accelerating the rate of universal expansion. As this acceleration continues, dark matter in space somehow increases proportionately; as spaces doubles, then dark matter doubles in mass and volume and causes more acceleration. Additionally, cosmologists have calculated that all galaxies should have flown apart by now because gravitational force and the distance between stars are not strong enough to keep galaxies together. The powerful force that keeps galaxies together is estimated to be dark matter. Christians believe that God is sustaining His creation (Colossians 1:16-17) through the use of dark matter.

Review: "God's invisible attributes, His eternal power and divine nature, have been clearly seen, being understood through what has been made," Romans 1:20. The fine-tuned universe testifies day to day of the glory of God. One of the oddities of the Big Bang theory is that on the first try, it perfectly laid the foundation and set in motion all the laws of physics, biology, astronomy, and chemistry. After violating the first and second laws of thermodynamics, then all the laws of physics were perfectly in place. Yet, the backbone of Darwinian evolution is a theory that suggests that random, unguided mutations eventually, after billions of tries, got it right. This is incongruous with the rest of evolutionary constructs.

Another law that the Big Bang hypothesis violates in physics is a principle law called the conservation of angular momentum. This states that the energy, linear momentum, and angular momentum of an isolated system all remain constant. Big Bang theorists contend that at the moment of the Big Bang, the singularity was spinning ultra fast. Thus, matter that came out of the Big Bang should have conserved the angular momentum of the singularity.

For example, if a bunch of kids were spinning on a really fast merry-go-round, and it spun so fast that the kids spun off, the kids would spin off with the same spin direction of the merry-go-round. Why? The outside portion of the kid's mass would be spinning faster than the inside portion of the kid's mass. The kids conserve the angular momentum of the merry-go-round. How does this work for the universe and the Big Bang hypothesis? Well, cosmologists theorize that at the moment of the singularity, it was spinning at a tremendously high velocity. This means that at the Big Bang, all the matter that spun off to form the galaxies would have maintained the same directional spin as the source. But this is not the case; some galaxies spin counterclockwise, some with a 90° tilt, and so on.

How do evolutionary cosmologists get around this? They hypothesize that matter bounced into each other and altered their spin. However, this is not exactly the vector of matter from an explosion. Upon any explosion, the outer portion of the explosion will have the highest velocity moving away from the center. Then the next layer in will have a slightly slower velocity moving matter away from the center and so forth. With each fragment or layer getting closer and closer to the center of the Big Bang, the matter will have a slower and slower velocity leaving the center. This means that as the initial fragments of matter (which would become the size of a galaxy) spun off the singularity at the Big Bang, those initial fragments would be traveling faster than the next set of fragments. The Cartesian system utilizes "X" for vertical coordinates, "Y" for horizontal, "Z" for depth, and "T" for time, and may help us understand this principle. During the Big Bang, fragments that came off simultaneously would not share the same X-Y-Z Cartesian coordinates because they would be exponentially spaced apart from each other as they receded away from the singularity. Therefore, it is less likely that the galactic fragments would have collided. Thus, the theory that the fragments bounced off each other to explain why the galaxies aren't spinning in the same direction and therefore explain why the Big Bang is not in violation of the conservation of angular momentum theory is a tenuous argument.

Some Big Bang theorists contend that there was no matter at the time of the Big Bang, and the formation of matter occurred through time as energy collided together. And eventually, the matter coalesced <u>over time</u> to form planets, stars, and galaxies. To have matter coalesce, there needs to be a rotation at the core of the matter. Coalescing matter like a hurricane coalesces moisture is an excellent explanation of how we got all the planets, moons, and stars. This theory is accepted by both creationists and evolutionist, though they differ on the amount of time it took to form celestial bodies. But in order to have the coalescing component, there needs to be angular momentum—rotation. And to argue that energy formed mass <u>over time</u>, removes the closeness concept of the singularity that is essential in order to force massless particles to collide. The CERN Hadron Collider demonstrates the amount of intelligence, energy, and order required to cause massless particles to collide together in close proximity. The further apart energy particles get from other energy particles, the less likely they could impact each other. Thus, the <u>over time</u> component actually undermines the hypothesis that the energy in the singularity formed matter.

Review: Big Bang theorists contend that the singularity was spinning ultra fast. Therefore, matter that came from the Big Bang should have conserved the angular momentum of the singularity. However, the galaxies and planets that we observe, do not spin in the same direction. Indicating that the spinning singularity hypothesis is incorrect.

NASA has suggested that the universe is not spherical, but is flat like a throw rug, or like a saddle on a horse. Some cosmologists suggest that this presents a problem for Big Bang theorists because this shape represent that the singularity had order, almost like designed order, at the beginning of time.

Why is a flat universe a problem? When objects explode in a vacuum without resistance, they explode in a spherical fashion, which means the universe should be spherical, not flat. In other words, there had to be some type of design involved in the singularity of the Big Bang so that when it occurred, it didn't obey a natural spherical explosion, but a flatter expansion. And any mention of a design of the singularity is not acceptable for their hypothesis. With an explosion, the place of the first displacement of matter would be at the weakest link, so even if they try to solve the flatter universe with an explosion following the weakest link at the equator and with the poles holding their integrity for a few seconds longer, it still means that the singularity had design, and that is a problem for Big Bang theorists.

When fireworks explode, they can cause the explosion to be spherical, circular, or in some other design, but this presents design. No flat-designed firework explosion occurs by a natural, explainable process. There is intelligence and design and a designer involved. Well, this leads back to a creation model and not random unguided events to bring about the order of the finely tuned universe.

Other cosmologists hypothesize that the flat universe that we perceive is just one morsel or thread of cluster galaxies, like a vast webbing of filaments, and our galaxy is in a tubular channel along with millions of other galaxies that make up one of trillions of tubular networks of connector threads or connector filaments. They suggest a bird's-eye view in which our galaxy is a nano-speck in one giant neural structure of a brain that we call the universe. If this theory is correct, then there is order and design in the positioning of each galaxy, as the galaxies are positioned together to form clusters in an elongated tubelike formation, and these channels of filaments connect other clusters of galaxies. The computer simulation models indicate lots of order, design, and formation. This is perfect for the creationist, but there should be randomness according to evolutionists.

Review: The Big Bang hypothesis violates observable evidence that the universe is flat because a natural explosion occurs spherically, not flatly. The notion that the universe is flat suggests that the universe was designed. Big Bang theorists will have to adjust their hypothesis again. Yet, the Bible remains unchanged.

Let cosmologists scramble to come up with an answer. Believers in the Word of God can relax. Why? The Bible may have clarified the solution 3,500 years ago. How?

When God created the universe, He <u>stretched</u> out the universe (Isaiah 40:22) in a way that revealed His design, and He told Moses to write about it. The words that Moses was inspired to write by the inspiration of the Holy Spirit were not fully understood by Moses. Gen. 1:6:

> Then God said, 'Let there be an <u>expanse</u> in the midst of the waters, and let it separate the waters from the waters.' God made the <u>expanse,</u> and separated the waters which were below the expanse from the waters which were above the expanse; and it was so. <u>God called the expanse</u>

heaven. And there was evening and there was morning, a second day.

This is the creation of our atmosphere, which was a direct result of God stretching out the universe on the second day. The KJV and NKJV use the word *firmament* for expanse. *Firmament* and *expanse* are synonyms and mean the same thing. The Hebrew word for "firmament/expanse" that God inspired Moses to use is *raqiya* (#7549), which means firmament or visible arch of the sky or expanse. Note that the visible arch of the sky potentially refers to the canopy of salt water that hovered around the atmosphere. *Raqiya* is derived from the root word raqa (#7554). The root word *raqa* means to pound the earth (as a sign of passion), by analogy to expand (by hammering), by implication to overlay (with thin sheets of metal), beat, make broad, spread abroad (forth, over, out, into plates), stamp, or stretch.

The inspired Moses wrote that when God began stretching out the universe, that God expanded it like a blacksmith expands metal to make a sheet of thin metal, with violent force, sparks flying, hot glowing metals, and steam from vaporized water. This is the Big Bang on the second day of creation.

We learn today that NASA has suggested that the universe is flat, which seems to fit with the Word of God and the notion that God expanded the universe like a blacksmith would hammer out thin sheets of metal from one lump. Mankind has just recently formulated the Big Bang theory, and creationists have resisted this notion, but the words Moses was inspired to write 3,500 years ago indicates a violent expansion with force, sparks, glowing hot magma, water, and steam, but on a cosmic level. All the things that would be associated with a blacksmith expanding matter. I was taught that all of the Big Bang was an evil concept, but in going to the root word of expanse, a clear picture emerges of the expanse on the second day of creation being a violent Big Bang (but not with the singularity premise, and with all matter already existing from its creation on the first day). And this reveals that the Word of God was supernaturally written because Moses could not have known that the atmosphere was flat and thin on the spherical earth, or that the universe contained flat portions or is flat entirely, or that the expansion was a violent force. Imagine the problem for creationists if Moses had written that God spherically expanded the universe with gentle separation; modern man would jump all over that error, but there is no error in the Bible, and no conflict with any of the sciences. It is ironic that modern science is just now catching up to the Scriptures and has revealed a mystery of why Moses was inspired to use *raqiya* (raqa is the root word) for expanse/firmament.

Review: The universe is flat or contains flat components to make up the whole, and the atmosphere is flat and thin around the earth, and God's word alludes to them as such 3,500 years before mankind even knew anything about the universe, providing more evidence of the divine nature of the Bible. The singularity of the Big Bang hypothesis violates the first law of thermodynamics (the conservation of energy), the second law of thermodynamics (entropy), the first law of motion, (the conservation of angular momentum), and the Word of God. The Bible is in harmony with the laws of physics, and they are in harmony with the Bible.

Group Discussion:

1. How did this chapter effect your faith in the Word of God?

2. Since Moses had no way of knowing what modern science has just determined, that the universe expanded violently, yet he used "raqyia" to accurately describe the expanse, what does this tell you about the divine nature of the Bible?

Chapter 28
Evolution versus Mathematics

Chemical evolution (abiogenesis) is the theory that all life stems from one single-celled life form, which evolved from nonliving material in a complex chemical pool of amino acids. Darwinian evolution is that everything in nature can be explained by the natural processes of survival of the fittest and random, unguided mutations. And from these natural processes, simple life forms gradually became more and more complex over time. There is no need for a supernatural creation event, and there is no need for God. We'll study evolution versus physics, chemistry, biology, mathematics, and the Bible.

First, let's look at evolution from a mathematical point of view. In basic math, any number (including infinity) multiplied by zero equals zero.

Addressing this evolutionary dilemma of violating Louis Pasteur's discovery that no living thing comes from nonliving material and that there is no spontaneous life, George Wald writes:

> The important point is that since the origin of life belongs in the category of at-least-once phenomena, time is on its side. However improbable we regard this event, or any of the steps which it involves, given enough time it will almost certainly happen at least once. . . Time is in fact the hero of the plot. The time with which we have to deal is of the order of two billion years. What we regard as impossible on the basis of human experience is meaningless here. <u>Given so much time, the "impossible" becomes possible, the possible probable, and the probable virtually certain. One has only to wait; time performs the miracles.</u> ("Innovation and Biology," 1958, in the *Scientific American*, Volume 199, p. 100.)

George Wald, elucidating the belief of the evolutionary community, violates a simple principle of mathematics. Zero times any number, no matter how large, still equals zero. But evolutionists adhere to this erroneous formula: nonliving material (zero) x long-enough time = life. But here's what evolutionists do to make their hypothesis seem viable: they add some important sounding words like *primordial* and *abiogenesis* and call the theory fact. And then ridicules anyone who objects to this leap of faith, and argues that abiogenesis is not evolution, yet each evolutionists believes it.

George Wald said, "One has only to wait; time performs the miracles." At least he admitted that there was a miracle. He just didn't believe that God performed the miracle; "time" performed the miracle instead. On a Biblical level, this violates the first law of the Ten Commandments, "Thou shalt have no other Gods before me." George Wald, speaking for all the atheistic evolutionists, is unwittingly saying that "time" is the god that performs the miracle. This is blasphemy. And guess what? All those who believe that "life came from nonliving material" are failing at the rudimentary level of science, and that is observability and testability. Since this is not observable and fails basic math, the hypothesis should be scrubbed, not dogmatically believed by blind faith.

Review: Evolution's spontaneous generation of life from nonliving material violates basic mathematics: zero x any number = zero, not life.

Let's go to an even simpler mathematical law to see how evolution violates math. Consider that any number plus zero equals the original number, represented by $n + 0 = n$.

Now, let's say "n" represent the primordial soup of complex chemicals and lightning and whatever condition evolutionists can come up with as the sum of nonliving materials.
And let's say "0" represents the absence of life. We come up with the equation: nonliving material +

absence of life = nonliving material.

Even if we wait and wait (which is "time" multiplied by "0"), the answer is still the same, (absence of life x 4.6 billion years) + nonliving material = nonliving material existing for 4.6 billion years. No matter how long one waits, the answer is still the same—no life.

Review: Evolution's notion of the spontaneous generation of life coming from nonliving material violates the most basic mathematical formula: $n + 0 = n$.

How do evolutionists get around this violation? Some have suggested that alien life initiated life on Earth. But we are still back to the same question—instead of who/what initiated life on earth, now it's who/what initiated life on the alien home world that initiated life on earth. The violation still continues, but now it's more in a mythological realm.

Some, such as Richard Dawkins, hypothesize that a meteor/asteroid (the alien) potentially brought just the right amount of amino acids to the planet. These building blocks then formed proteins and ultimately DNA in the primordial pool and initiated life.

Either way, this is still a violation of the simple mathematics of something from nothing: life from the nonliving. Evolutionary scientist are earnestly trying to create life from non-living material. Although every attempt has failed. Let's venture into the realm of fantasy for a second. Let's say that mankind did successfully create life from nonliving material. Would that solve the problem? That would validate the notion that a creator was required and behind the creation of life, so it would not be chance. It would be mankind making life in the image of God and acting as God and creating life. Mankind would then be the intelligent designer of the creation. It's a no-win situation for spontaneous life believers. Either they fail to prove life can come from nonliving material, or they are successful, and they authenticate that a creator is required.

Do I think mankind will be able to create life via lab testing? Yes. There is a probability that mankind will be successful in the near future. Researchers are already working on isolating mutated recessive DNA codes that lead to diseases; this is the tip of the iceberg of what mankind will be able to do. At this pace of accelerated knowledge and technology, whatever mankind can dream up will be tomorrow's reality. Why? Gen. 11:1–9, in reference to the Tower of Babel says:

> Now the whole earth used the same language and the same words . . . They said to one another . . . "Come, let us . . . make for ourselves a name, otherwise we will be scattered abroad over the face of the whole earth." The LORD came down to see the city and the tower which the sons of men had built. The LORD said, "<u>Behold, they are one people, and they all have the same language. And this is what they began to do, and **now nothing which they purpose to do will be impossible for them**</u>. Come, let Us go down and there confuse their language, so that they will not understand one another's speech." So the LORD scattered them abroad from there over the face of the whole earth; and they stopped building the city.

It seems we are approaching a similar point in the history of mankind again. Mankind has returned to the point of being of one voice, and it seems that humanity is saying there is no God. Mankind has returned to the point of using the same words, and those words seem to say that we evolved from nonliving material through evolution. Mankind has returned to the point of being one-minded by using resources and money in an attempt to prove evolution and disprove the Bible and therefore disprove God. God said that when the people are one, "Nothing which they purpose to do will be impossible for them." Therefore, mankind is right now at the point of being successful in whatever endeavor it chooses to embark upon. There is evidence in Strasbourg France, where people have built what seems to be a replica of the Tower of Babel, with production posters that read, "Many tongues, one voice."

Mankind is again at the precipice of repeating their actions of the Tower of Babel, that is, of making a name for themselves and trying to take away from the name of God. But this will only repeat the same result and bring the return of the Lord with His judgment.

So evolutionists are in a no-win situation. Their attempt to prove there is no God and make a name for themselves is only storing up wrath for themselves and will only fulfill Scripture and bring God down to Earth with His judgment as it did for the people of the Tower of Babel.

Wise sages have always said that if people don't remember history, they will repeat it. Look at the parallels: At the Tower of Babel, mankind had one voice and one language and used the same words; people pooled their financial resources to make a name for themselves and take away the name of God. They brought judgment and brought God to come down to the earth. We are at the same moment of historical precedence, just ~4,000 years later. Mankind has built such facilities as the CERN Hadron Collider to find proof of the "God" particle that started everything and that there is no need for God to perform the miracle of the Big Bang. Mankind has spent billions of dollars on searching the stars, not to marvel at God's creation, but to find life on other planets to explain via natural means how life began on earth, without God to start it all. This is a repeat of the same moment in history, the same thinking that mankind had at the Tower of Babel. And the result will be the same: judgment, the fulfillment of all the "end time" prophesies and the return of God in the flesh upon the earth, Jesus.

The promise of God is that through the evil of mankind, God will make good come out of it for those who love Him (Romans 8:28: "And we know that God causes all things to work together for good to those who love God"). This is best exemplified with the crucifixion of Jesus (Act 2:23–24) and best summarized with this verse: "You meant evil against me, but God meant it for good in order to bring about this present result, to preserve many people alive" (Gen. 50:20).

Review: Evolution violates the simple laws of math that say that zero times any number equals zero and that any number plus zero equals the original number. Mankind is on the verge of repeating the same mistake that humans made at the Tower of Babel, and the same outcome awaits mankind.

Does the Bible violate the same mathematics when God created Adam and Eve? Let's take a look. God said (Gen. 1:26–27): "Let Us make man in Our image, according to Our likeness . . . So God created man in His own image; in the image of God He created him; male and female He created them." Then God gives us more detail of the creation of man with Gen. 2:7: "And the LORD God formed man of the dust of the ground, and breathed into his nostrils the breath of life; and man became a living soul." Did God violate mathematics and the law that says that anything multiplied by zero still equals zero? Let's break down the key words of the last bit of detailed information:

Formed (#3335): *yatsar*: to press, squeeze into shape, to mold into form, as a potter.
Dust (#6083): *aphar*: dust, clay, earth, or mud.
Breathed (#5301): *naphach*: to puff, to inflate, or blow hard; expiration.
Breath (#5397): *neshamah*: **breath,** a puff, as in the wind, angry or vital *breath*, divine inspiration, intellect. or an animal, (that) breath (-eth), inspiration, soul, or spirit.
Life **(#2416):** *chay*: alive, (as a noun) life, or living thing.
Living **(#2416):** *chay*: alive, (as a noun), life, or living thing
Soul (#5315): *nephesh*: a **breath**ing creature, i.e. animal or vitality, bodily or mental, soul, mind, desire, heart, man, him, her, me, one, own, person, self, or will.

A salient point is the similarities between *life* and *living*. Both words are defined by Strong's Concordance (#2416) as *chay*: alive, (as a noun), life, or living thing. This clarifies that from the Living thing, man became a living thing, or from the Life, man became a life. This clearly sums up that man didn't come from nonliving material, but man/Adam came from living life, from God Himself. This means life begot life, not that nonliving material begot life as in the evolutionary model. This leads to the next part, which is what became alive? And that is the soul, the breath of man. What has always been the source of the soul and the breath of life? Well, that would be the Breath of God—the Holy Spirit. There are similarities between the definitions of *ne<u>s</u>hamah/Breath* (#5397) and *nep<u>h</u>esh/Soul* (#5315), and the root words they are derived from are similar as well. Both definitions share the meanings of *breath, mind, vital, and soul*. Thus, we may infer that the Breath of God begot the breath of man, and the Soul of God begot our soul. The Bible is in harmony with science again, that our mind, breath, vitality, consciousness, and soul, came from God's mind, breath, vitality, consciousness, and soul. Evolution stands alone in contrast to biological and medical science, with the hypothesis that life came from nonliving material. Science affirms that no living thing can come from nonliving material.

How did God make up the *life* that made man, shaped from red clay, *alive*? This is splitting the Trinitarian hair, but here it goes. When God (Elohim = plural form of the one God, i.e., Father, Son, and Holy Spirit) created everything, He spoke everything into existence. The breath that came out of God's mouth as He was speaking was the moving of the Holy Spirit, and the Holy Spirit entered the man, who was shaped from red clay, and from L*ife,* Adam became *alive.* How do we know of the Trinity that this was the Holy Spirit? When God breathed the "Breath of Life," the word for breath was *neshamah* (#5397); that is, <u>wind, angry/vital breath, divine inspiration, divine intellect, soul, or spirit</u>. The Hebrew word for *spirit* (as in the Spirit of God moving over the surface of the waters; Gen. 1:2) is *ruwach* (#7307), which means <u>wind, breath, violent exhalation, or spirit</u>. They have very similar definitions. Ruwach is closely associated with *Neshamah/Breath*. In fact, *ruwach* is sometimes used as the Holy Spirit (Gen. 1:2) and sometimes used as breath (Gen. 6:17). Therefore, when God shaped man and then spoke him into being alive, it was because God the Spirit, the breath of Life, moved and entered Adam and made him alive. Life begot life.

Review: The Bible indicates that the Breath of God begot the breath of man, and the Soul of God begot the soul of man, and the life of God begot the life of man. All that have the breath of life came from a living God. Only evolution stands firm in direct violation of biological and medical science that life cannot come from nonliving material.

Does mankind consists of two parts, such as a dichotomy—body and soul/spirit, or three parts, such as a trichotomy—body, soul, and spirit? The reason this is challenging is because the spirit of mankind came from the Holy Spirit, and the soul of man came from the Breath of God—the Holy Spirit. Both the soul and spirit of man came from the Holy Spirit. Therefore, there is some overlap in the definitions and, subsequently, some difficulties discerning between the soul and the spirit. Whether an individual accepts a dichotomy or a trichotomy interpretation depends on one's personal perspective. For example, it is true that upon being conceived, when the soul is fashioned and the spirit is joined to the body, then the soul is never separated from the spirit going forward. Therefore, the dichotomists are correct that there is a material aspect to man—the body—and an immaterial aspect to man—the soul and spirit. However, there is more information to consider.

Medical science details the difference between the articulation of two bones called a joint, versus the marrow within the bone that produces blood. And we have all tried to explain the difference between our thoughts and our intentions when being misunderstood. For example, when someone gets their feelings hurt because of our words, they may say something like, "I am so sorry it came across that way; my intention was not to hurt you, I thought I was helping you." The thought was good, but the intended result didn't work out. The examples illustrate that it takes a specialist to know the difference between closely associated terms. It is difficult to discern between the two because there is a fine line between them. But God always knows whether our intentions are good or bad, He always knows the line between the joint and the marrow and between our thoughts and intentions. In fact, Hebrews 4:12–13 says, "The Word of God is living and active and sharper than any two-edged sword, and piercing as far as the division of soul and spirit, of both joints and marrow, and able to judge the thoughts and intentions of the heart. And there is no creature hidden from His sight, but all things are open and laid bare to the eyes of Him." Therefore, there is a division between the soul and the spirit of every creature. Know this, that the Bible never wastes words, so if the Bible uses the word *soul* in some locations and uses *spirit* in other locations, then there is a reason why. We may not fully know why for each verse, but that doesn't negate that the two closely associated, albeit different words, soul and spirit, have different applications and meanings.

Considering the above and the notions that the spirits of all life was created on the first day of creation and the soul/breath of life, was made on the sixth day of creation, then there seems to be a definable difference. The bird's-eye view of Scripture suggests that the spirit of mankind connects the body and soul to the spiritual realm. Ephesians 6:10–20 elucidates the point, "Our struggle is not against flesh and blood, but against . . . spiritual forces of wickedness in the heavenly places." The flesh obviously is the body, and since God doesn't waste words, then the blood adds additional information. The blood carries oxygen, which is from our breath, and our breath comes from the breath of life, which made us a living soul (Gen. 2:7), then the blood represents the soul. Thus, we may interpret Ephesians 6:12 to mean that we do not wrestle against human bodies and human souls, but against spirits in the spiritual realm. Therefore, our spirit connects our soul and body to the spiritual realm. Our spirit is either of the Light—saved, and this causes our soul and body to bear Godly thoughts and actions, or our spirit is of the darkness—not saved, and this causes our soul and body to not bear Godly thoughts and actions. We have no power over the spiritual realm by our own merits. This is where the Bible tells us that God is our source for power, faith, truth, wisdom, knowledge, and understanding, and through God, we have power over the spiritual realm. Our spirit may be considered our subconscious. There is nothing we can do to work on or improve this area by our own merit, but only by obeying—it is a gift from God by His grace alone. This is why salvation is exclusively by God alone, and no work can earn or improve on salvation (Ephesians 2:8).

The soul (a breathing, conscious life) of mankind seems to connect the body to the spirit. Spiritually speaking, without the Holy Spirit, we are dead in our sins though still breathing, and our soul will only desire for our body to carry out self glory, self worship, greed, and only evil, and we are alienated and hostile in mind toward God (Col. 1:21–22). And with the Holy Spirit, we are alive in Christ, and our soul desires for our body to be morally obedient to God, to submit to the authority and ownership of God, and to revere God. This is where the Bible tells us to put effort in, be diligent, seek truth, repent, and read the Scriptures to grow in wisdom, knowledge, and understanding. And we are to be sober minded, obedient, holy, and so forth. This is our part in doing work that authenticates that our spirit is saved (James 2:14–26). Physically speaking, without the Breath of Life, we are dead, and dust we shall become. And with the Breath of Life, we are alive to think, act, breath, have desires, and so forth. This is where the boy Jesus studied the Scriptures and increased in wisdom (Luke 2:52).

The body of mankind seems to connect the soul and spirit to the physical realm. The physical body manifests and represents the condition of the soul and spirit that dwells within the body. The body is a temple (I Corinthians 6:19), and either God dwells there and we see that manifested in our actions and words, or it is a temple with either no spirit dwelling within or an evil spirit dwelling within, and we see the lack of Godly actions and words. Note that God purchased every body (temple) with His blood, but the free will of the soul must accept God through faith alone.

Review: The Scriptures suggest that life is made up of a body, soul, and spirit—a triune human made in the same image of a triune God (Gen. 1:26).

Now back to the question: Did God violate mathematics and the law that anything multiplied by zero still equals zero? No, because God breathed the breath of Life into Adam's nostrils from His own essence, and man became a living being. Adam became a living soul with a spirit from God the Spirit, as God the Father gave God the Son the Words to speak (three aspects of one God). Essentially, man was created from the very essence or a part of God. Not only were we created to look like God, and not only were we created to have some of God's characteristics (Gen. 1:26), but God gave us the breath of life from His very being (Gen. 2:7). Therefore, the Bible doesn't say that Adam came from nonliving material—no, quite the contrary—the Bible says Adam came from "The Life Source," from God's breath (breath = ruwach = Holy Spirit) of Life. Therefore, God breathed an essential essence of Himself into man to give man a soul and spirit. Man came from a living God. Life begot life.

The notion that Life begot life is supported in the genealogy of Luke 3:23–38: "Jesus the son of . . . the son of David . . . the son of Jacob, the son of Isaac, the son of Abraham . . . the son of <u>Adam, the son of God</u>." Adam was the first son of God. Life begot life. So the formula looks like this:

God spoke x (nonliving material red clay shaped into the form of man + Holy Spirit) = living man with the Holy Spirit dwelling inside man = God spoke x (0 + Breath of Life) = life.

Adam was the first son of God. Life begot life. This doesn't contradict the statement, "Jesus is the only begotten Son of God" (John 3:16) because Jesus has the title of being "First Born" (Colossians 1:15), but Jesus has always eternally existed (Hebrews 13:8 "Jesus is the same yesterday, today, and forever") as the co-creator of all things (John 1:1–3, Gen. 1:1, Colossians 1:15–17). This is rectified by understanding Old Testament, Abrahamic, and Davidic ways. And the title of "First Born" usually goes to the first born son who was physically and literally born first. But there are instances where the title went to a younger son, such as with Isaac (Gen. 21:13), Jacob (Gen. 25–27), King David (I Samuel 16:10–11 and Psalm 89:27), and even the kingdom of Israel (Exodus 4:22). They were all given the title of "first born," but they weren't physically born first. Isaac was younger than Ishmael, Jacob was younger than Esau, King David was younger than seven other brothers, and there were many kingdoms on earth before God declared Israel His "first born."

When Jesus was born in Bethlehem 4,000 years after Adam, it had nothing to do with when Jesus came into existence. For Jesus has eternally existed as God in the flesh. However, Jesus shed His immortal flesh and put on mortal flesh and literally and physically was the offspring of God the Father, as was Adam. That's why Colossians 1:15 declares Jesus the first born of all creation; this is saying that Jesus has the title of being the First Born of God the Father physically, and spiritually over the real first human born of God, Adam. Jesus has all the inheritances from the Father belong to Jesus.

The change from the immortal flesh of God the Son to the mortal flesh of God the Son (yet still maintaining His immortal soul and spirit inside), through the birth of the virgin Mary, was for the purpose of fulfilling God's Word with the final blood atoning sacrifice for all of Elohim's creation—not so that Jesus could earn the new title of first born. Jesus already had the eternal title "First Born" in the spiritual realm before Jesus in Gen. 1:1 co-created everything with His Father and the Spirit. Now, at Jesus' physical birth, He revealed that He has always been the worthy possessor of the title of "First Born" of all of creation over the physical realm as well as over the spiritual realm.

Don't let anyone try to convince you that Jesus was a created being because He is physically and spiritually the possessor of the title "First Born" of all creation. It's simple; one can't be created when you are the creator of all things (John 1:1–3 and Colossians 1:15–17).

Review: The Bible is in harmony with mathematics, as man was created from Life. Life passed on life. Life begot life. That's why Adam was the first born of God, but Jesus is the worthy possessor of the title "Firstborn of all creation" because all things were made through Him.

God has finished creating all things (John 1:3) and is now sustaining His creation (Colossians 1:17). And all the spirits of every human and every moving thing that has the breath of life and angelic hosts were created during the six days of creation. I contend and will show later that our spirits began on the first day of creation. When a baby is conceived, the spirit is not created. This means that with each conception of a life/soul, the spirit is not created, it is united with the energy carrying "unit" passed down by the parents, called the body, and becomes a living soul. And at conception, the spirit joins the newly formed fertilized egg in the womb and becomes a living soul when the egg attaches to the endometrium of the uterus to receive the blood; the life. That's why abortion is murder and continues on the pagan worship of the Ammonite god, Molech (Lev. 18:21, Jeremiah 32:35). That's why a new birth is God sustaining His creation, not still creating, as He did on the first six days of creation.

The body, soul, and spirit are from God's essence; literally, Adam was comprised of pieces of God, if you will. The spirit of man is one of the things created with the light on the first day without the sun, moon, and stars being created until the fourth day of creation. God is light, and He took some essence of Himself to create the spirits. And that is why the souls and spirits and angelic hosts are immortal because God is immortal. So essentially, Life begot life, God begot the souls and spirits from Himself. The formula for the creation of Adam would be as follows:

Shaped man of the red clay, but not alive = 0.
God = Infinite = ∞.
God's Spirit/God breathed = infinite = ∞.
  (Uncreated man + God's Spirit) x God spoke = Adam = $(0 + \infty) \times \infty$ = Adam.

Now according to this formula, this would make a portion of Adam infinite, or in other words, immortal, and a portion of Adam mortal. And that is exactly what mankind is, a mortal temporal body with an immortal soul and immortal spirit. The mortal aspect of our body is exemplified with the paraphrase, *from the dust you came and from the dust you shall go* (Gen. 3:19 and Ecclesiastes 3:20). Eventually, humans will be clothed by God with an immortal body to house our immortal soul and spirit, but currently we have a mortal body. When everyone gets their immortal body, the source will be divine, and some will worship God with that eternal immortal body, but most will suffer in hell with that eternal immortal body.

The Bible says, "Jesus is the Way, the Truth and the Life" (John 14:6). All of those that are not with Jesus are dead in their sins. Is John 14:6 only referencing the spiritual aspect (as far as salvation), suggesting that Jesus is the Life for salvation only, and not also physical life? Maybe, but I'm thinking it also represents the physical. How about John 1:3–4: "All things were made through Him (Jesus), and without Him nothing was made that was made. In Him was life." Jesus is the "life" giver in that Jesus was "life" from the beginning. Thus, as Jesus spoke, He breathed His "life" force, the Holy Spirit, went into Adam, as the Father gave Jesus the Words to speak creation into existence. Therefore, Elohim (Father, Son, and Spirit) created everything.

Review: The Bible is in harmony with mathematics because the life found in the first man was from a living God. The Bible explains that life on earth came from Divine Life and that life begot life, which is in harmony with math. However, evolution is in violation of math by suggesting that life came from nonliving material.

What about all the animals? Does the Bible indicate how they were created, and does the Bible violate the same math that evolution's spontaneous life violates? Let's take a closer look. The Bible gives us some general information about the creation of all matter in the universe, though void of life and formless, with Gen. 1:1: "In the beginning [time] God created the heavens [space] and the earth [matter]." Then God gives us some detail with Gen. 1:20: "Then God said, 'Let the waters abound with an abundance of living creatures, and let birds fly above the earth across the face of the expanse of the heavens.'" So God created great sea creatures and every living thing that moves, with which the waters abounded, according to their kind, and every winged bird according to its kind. And verse 24: "Then God said, 'Let the earth bring forth the living creature according to its kind; cattle and creeping thing and beast of the earth.'" Then we get a little more detail about the creation of the animal kingdom with Gen. 2:19: "**Out of the ground the LORD God formed every** beast of the field and every bird of the air." Then we get some more clarity with Leviticus 17:10: "And whatever man . . . who eats any blood, I will set My face against that person who eats blood . . . For the life of the flesh is in the blood, and I have given it to you upon the alter to make atonement for your soul, for it is the blood that makes atonement for the soul." And then finally Gen. 6:17: "I am bringing the flood of water upon the earth, to destroy all flesh in which is the **breath of life**."

    There is a lot of information here; the first for this section is about everything that has the blood and breath of life on earth—the creatures became living souls: (a) God formed them out of the ground just like God formed man; (b) and God breathed the breath of life in them, just like God breathed the breath of life in man. The same concept applies, that everything on earth that has the blood of life in them came from the breath of Life, a living God. Life came from Life. The Bible is consistently in harmony with mathematics, even with the creation of the animals. This is why all creatures on earth share some similar DNA. All that have the blood of life in them originally came about via the same breath of Life (Holy Spirit) entering their oldest ancestor. And this is why the Levitical law required life for life, breath for breath, soul for soul. This law in Leviticus was for governments to rule the people, not for people to exact their own revenge. Jesus clarified to turn the cheek, and allow God and the government to render justice according to the crime.

    Some naysayers may argue that the text doesn't directly say that God breathed the breath of life in the animals. But, no one can talk without the breath of life flowing through them. Try talking without any breath; it can't be done. We were created in His likeness. Just imagine God saying, "Let there be," with His voice having the "Breath of Life" (Holy Spirit) flowing through Him. With His Word, He created everything. His words are powerful, and since we are an essence of Him, our words have power. With our words, we confess God as Lord of our lives unto salvation (Romans 10:9); with our mouth, we can edify our brethren or cut down our brethren (James 3).

    The Scientific Method requires a hypothesis to be observable and testable. 100% of the time we observe life always comes from a prior life. And 100% of the time we test that life always comes from a prior life. Therefore, there are only two logical conclusions. Either this concept of life from a prior life repeats for eternity past, which cannot be because we know the earth is finite and had a beginning. Or at some point in time in the past, an immortal life begot life on Earth. Those are the only two options from what we observe and test. A hypothesis that must be fully rejected based on the scientific method, is that at some time in the past life came from non-living material because that has never been observed or established by tests. And for this reason, only the Bible is in harmony with science. And only the hypothesis that evolutionists start with is a violation of science. And by the way, the very same scientific method applied to the origin of the DNA results in the same conclusion. Always, 100% of the time, DNA for life comes from a prior life. Never has it been observed or tested that the DNA for life has come from non-living material. Therefore, the study of modern medical science of genetics proves the Bible is the only one in harmony with science, and proves that only evolution is in violation with the science of genetics.

    To split the Trinitarian hair; Jesus is the one doing the talking (Gen. 1:26, Colossians 1:15–17, John 1:1–3). The Father gives all the Words for Jesus to speak into existence (John 12:49), and the Holy Spirit is the breath that moves and enters the newly formed being to begin life (Gen. 2:7 and Gen.

1:2). All three aspects of the one God were equally involved in creation as one God—just like a human being has three aspects to make up one person: body (Jesus) that is the image of the mind, the mind/soul (Father) that gives the body the words to speak and connects the body to the spirit, and the spirit (Holy Spirit) that connects the mind and body to the spiritual realm. That is why, in Gen. 1:1, "God" in Hebrew is "Elohim," which is the plural form of the singular God, the Trinity. This is also the same reason that Gen. 1:26 has God saying Let Us make man in Our image according to Our likeness. God is using plural personal pronouns to describe God (The Father, Son, and Holy Spirit) as one.

It is sad to hear about believers in God, who reject what God said He did at creation to form life and believe the schemes of man on how life began. Man's way, "something from nothing," is proven to be a violation of mathematics, biology, and medical science. God's way has life coming from Life, and since God is infinite and all powerful, then "something came from everything." The two views are polar opposites. Don't think you can believe a little of both. It's either all of one or all of the other. There is no in-between position, no lukewarm spot, no partial scatterers, and no partial gatherers; you are either all for the Word or all against the Word. One can't even say, "This isn't my fight; I'm not involved." There is simply no middle ground (Matthew 12:30).

Review: Evolution's spontaneous life from nonliving material violates the most basic mathematical formula (zero x any number = zero). The Bible is in harmony with mathematics because the life found in the first man was from a living God. So the Bible explains that life on earth came from life. God begot Adam (Luke 3:38). Life begot life with the formula of God x (nonliving material + Holy Spirit/*chay*/breath) = life. Evolution and creation are polar opposite views. Evolutionists believe something came from nothing, while creationists believe something came from everything.

We have established the idea that evolution violates the simplest formulas of mathematics. What about other levels of mathematics, such as odds? Evolutionists believe that unguided, random mutations made meaningful changes to the DNA code that provided new information in order to evolve life into greater complexities and completely new kinds of creatures. What are the odds of evolution? What the general public doesn't know is that the human DNA code is so complex that if the DNA code was typed out on paper on the front and back sides and the pages were stacked on top of each other, the pile of pages would be as tall as the Washington Monument. Displace one small section of coding and malfunctioning mutations would occur, including disease, impairment, stillbirths, early death, or sterility.

Any computer programer will tell you that when writing software programs to perform a simple function, if there is one error, the program will not work properly. And computer software programs are similar to DNA coding in that errors in both result in malfunctions of the program. Both systems require a specific design to be able to function, and if left alone, will not add extra code. But evolutionists believe that random, unguided, mutations made meaningful changes in code to create new coding that wasn't there before for new functions or a new kind of creature. The odds of random, unguided mutations producing meaningful changes to proteins that are not life threatening to an organism—let alone producing a series of proteins and their accompanying genes and let alone producing a whole section of DNA coding having meaningful changes of kind—are beyond the threshold of possibility. Any computer programer will tell you that random, unguided changes to coding are more likely to degrade the code rather than enhance the code. And a computer programmer writes simple programs compared to the DNA code. Yet, the simplest living organism is trillions of times more complex than a computer program. To know that a simple computer program fails with one mistake in software coding, yet still believe that random, unguided mutations to a complex organism creates meaningful changes and leads to a new kind of creature, is an amazing leap of faith.

As a doctor, I can tell you that mutations accelerate the death process; they don't enhance life. And when a mutation occurs and kills the host, there is no information passed on to a surviving neighbor to tell them not to mutate a protein of the DNA coding in the same area. Usually, the body dies from a mutation; most of the time, the body rejects the mutation and wars against it. Sometimes, the mutation causes the body to be sterile (the inability to reproduce). Most of the time, mutated creatures are rejected by the opposite gender as being unworthy to reproduce with. Sometimes, a mutated creature is rejected and killed by the same gender. Usually, the harshness of the weather kills off mutated creatures. The mutated creature is always weaker and more prone to sickness and poorer health. If the weather doesn't kill off all the mutated creatures, then the predators get them. Mutated creatures are always slower and less coordinated, with less endurance and less strength. As a result, they are easy prey for predators and are the first to be killed.

Here are the obstacles for a mutated creature to overcome:
1. Mutated creatures are often born dead.
2. Mutated creatures often die shortly after birth.
3. Mutated cells accelerates death (e.g., cancer).
4. The body rejects mutated cells and wars against them; this leads to frequent sickness.
5. Mutated creatures are often sterile and are unable to pass on any inherited modified DNA coding.
6. Mutated creatures are rejected by the opposite gender and therefore can't pass on DNA.
7. Mutated creatures are rejected by creatures of their own genders as unworthy members of society and killed.
8. Weather kills off most mutated creatures because mutated creatures are weak and frail and prone to sickness and disease.
9. Predators kill off the remaining mutated creatures. Mutations make a creature slower or weaker, with less endurance and less strength; sometimes mutated creatures have all these deficiencies.

An even greater problem for natural selection proving evolution is that natural selection is based on a preexisting DNA code; one can't argue for evolution based on natural selection when the DNA code built into each creature for natural selection has no origin, but just an existence.

When one takes the sum of the problems that mutations have to overcome to evolve a new function or to evolve from one kind of creature to another kind of creature, it becomes an impossible feat. The probability for survival is such a small number that it renders the hypothesis implausible. MIT graduates can calculate the odds to be impossible, and computer programmers can tell you that random mutations ruin a program, not enhance it. But the bottom line is that if people take a bird's-eye view of the problem mutations have to overcome, they will see that the evolutionary hypothesis of random, unguided mutations being able to beneficially change a function of a protein or change the kind of creature is implausible—just based on the survivability of mutated creatures. And this doesn't even consider that there is no information in the DNA code to establish a meaningful change of function or change of kind. In other words, without the DNA information already being in the code, there is no way to get the code in without an intelligent designer putting it there. And the hypothesis of utilizing natural selection to pass on genes that have undergone random, unguided mutations turns a blind eye to the probability of the creature being sterile, stillborn, weak, and so on.

Mathematically, evolution is impossible. For when a mutation occurs, there are too many obstacles to overcome. And the altered DNA would still have no information for creating a new kind of creature; all that the altered DNA would have is the degraded DNA code of the original host creature. So as much as evolutionists want to believe that mutations build on other mutations to eventually generate a wholly new kind of creature, it is implausible. And don't confuse mutations with adaptations; they are different.

A mutation is a change from a normal cell to an altered cell by inherited trait or biochemical change in the genetic material, with abnormal function and/or impaired function. This change is not hardwired into the DNA, and the cells that mutate are foreign to the body. This causes the host to suffer on some level. These changes are random and unguided. Any unguided change to the DNA code produces an adverse effect, not a beneficial effect. In addition, any altered cells from an outside source still doesn't have the DNA information to utilize the alteration and build upon it. Mankind will soon be able to make guided changes to the DNA code to produce a beneficial effect, such as removing a mutated disease code in the DNA. This however, establishes that intelligence is required to change the DNA code, not unintelligent, random, unguided mutations.

Adaptation or speciation is a small change(s) to conform to a new environment. The ability to adapt is already hardwired into the DNA, and the cells that adapt remain normal to the body, and the DNA remains the same. Why? In the words of Grady McMurtry (Nine great proofs for evolution and why they are all false): "The Laws of genetics are conservative, not creative; these laws only allow for the copying or rearranging of previously existing information which is then passed on." Take for example the Galapagos finches; they already had the information in their DNA from their ancestors for a particular beak shape, and if their beak shape better suited the environment, then they would thrive and reproduce more offspring that also have the information in the DNA for the beak that allows them to thrive in a particular environment. Thus, the DNA information for that particular beak dominates the population through a process called natural selection. But this is based on information already existing in the DNA code, not new information embedded in the DNA code. This is not evolution—this is adaptation.

There are ways for DNA to change, such as combining chromosomes from parents. This is how a kind of creature will adapt over time to form different species of the same kind, and this is all based on existing information in the DNA, but the kind of creature will never change. This is how the wolf adapted through many generations to be related to the common house dog through a process of natural and artificial selection, but always based on exiting information in the DNA. This is a process called adaptation, not evolution. Another way of changing the DNA or genome is with random, unguided mutations from external stimuli, such as chemicals (toxic chemicals, drugs, etc.) and high-frequency energy (UV rays, X-rays, Gamma rays, etc.), and this always results in a loss of functional DNA information, which results in either impairment of function or a loss of function; never ever does a mutated DNA result in the gain of new functional information that results in new function or a new kind. Only in the science fiction movies will chemical or solar energy result in an improvement of the DNA. This is evolution. And this is the error of evolutionists; they believe that adaptation is evolution, but the two concepts are vastly different. Another means of changing the DNA code is with internal stimuli. Such things as stress, obesity, and other sins may adversely alter our genome, with the spiritual realm affecting the physical realm. In Exodus 20, the Ten Commandments, God says He will visit the iniquities of the fathers to the third and fourth generations of those who hate God. Did you know that sin adversely alters our genomes, and when we reproduce, our chromosomes may contain the mutated genome that resulted from sin. Thus, a sinful parent may pass down defective genomes in their chromosomes for fertilization, which passes on the affected genome from our sin. For example, say a father abuses alcohol, and on a spiritual level, he is worshiping himself with his lust for physical satisfaction and adversely affects his genomes for the internal organs. Now when he has children and passes on his mutated genome, then his heirs have a higher propensity of organ failure, therefore, fulfilling the Ten Commandments of having no other god before God, or else God will visit the iniquity of the father to the third and fourth generation. In this example, this resulted in a loss of efficient information in the DNA code. On a genetic level, a further deviation away from Adam's DNA, and since Adam's DNA 100% came from God's DNA, then the sin of the father, altered his DNA further away from a copy of God's DNA. And by the way, this example can be illustrated with all the acts of sin that may adversely affect our genome. Even stress can adversely affect our genome, including stress as a result of not trusting in God. When one trusts in God, they will have no fear though they walk through the valley of the shadow of death (Psalm 23). And when someone trusts in the Lord, they will receive a peace that surpasses all understanding (Philippians 4:6–7). However, God is merciful and full of grace. There is a way for offspring to break this curse from sin, and that is with forgiveness of sin and obediently following the precepts in the Bible. Forgiveness of sin and obedience to God cause the spiritual realm to beneficially heal the physical realm. With the body being free of a sinful lifestyle that adversely affects the genome, the body can repair the damaged genome with protein markers that are designed by God and embedded in the DNA to hunt down mutations and repair them. This is adaptation, not evolution because the information already exists in the DNA. This is why the Ten Commandments follows the warning to the fathers with; "but showing lovingkindness to thousands, to those who love Me and keep My commandments" (Exodus 20:5–6).

The adaptation aspect of genetics has limits to how far and to what extent life can adapt. It allows a closer copy of the original DNA to survive and be passed down. Changes to the genome are designed and guided by DNA coding. Evolutionists see adaptation and mistake this for evolution. But that is an error in deduction, as the ability to adapt is already in the DNA code, so it's not the DNA that changes, it's the body's response to the outside stimulus that is modified because the preexisting DNA tells it to.

Evolutionists often show examples of creatures that have lost a function because of mutations and proclaiming that this proves evolution and that mutations gain new information over time. For example, they cite flies with four wings and say, "See, evolution." But what they don't tell you is that the fly cannot fly, so it is an example of a lost function; other flies won't mate with it, so no genes get passed on, and sometimes they are sterile.

Stephen Meyer, author of *Darwin's Doubt* and *Signature in the Cell*, has written overwhelming evidence regarding the improbability of random mutations evolving new genetic DNA to form new functions or new kinds of creatures and about intelligent design in the DNA code. He has utilized the scientific method and a peer-review process to authenticate the notion that all life has intelligent design, and therefore an intelligent designer brought forth life.

Review: The kind of creature will not change with adaptation (governed by preexisting DNA); only a different aspect of the DNA code is predominately displayed via natural selection. DNA will change with mutations from sin, chemical, or high-energy radiation, and this always results in a loss of functional information that results in an impaired function or loss of function, and natural selection (governed by preexisting DNA) results in mutated creatures not being selected for mating. Random, unguided mutations are more likely to degrade the DNA than enhance the DNA and result in a different kind of creature. The odds of mutated creatures surviving and procreating to pass on their altered DNA code has too much to overcome: stillbirth, very short life span, sickness, weakness, malfunctioning vital organs, reduced immunity, sterility, rejection by the opposite sex, rejection by the same gender, weather, and predators. The odds are against evolution, and this demonstrates the extent of evolutionists' faith.

Group Discussion:

1. What are the implications for your life knowing that only the Bible, according to the scientific method; of observation and testing, is in harmony with biology and genetics, that is that life comes from a prior life, and the genetics for life also comes from a prior life?

2. How has this chapter strengthened your faith in the Bible and God?

Chapter 29
Evolution Versus Physics

Evolutionists believe that billions of years ago, when the earth was hotter, rain came down upon the earth and formed pools of complex chemicals of amino acids, the building blocks of life. Then, given the right conditions and enough time, life came from nonliving material and that the first life form was simple, a single cellular organism. Via the means of random, unguided mutations, some form of a proto-RNA replicating entity evolved into the first prokaryote life form, and the complexities of life that we see today came via a natural selection process called the survival of the fittest and random, unguided mutations.

The hypothesis of evolution violates the second law of thermodynamics (entropy). Entropy suggests that all things go from order to disorder, from complex to simple. Going from a simple single cell to the complexities today is in violation of entropy. The evolutionary counterargument is that outside forces applied energy to allow the smaller system to tend toward more order. And this is a valid argument on the surface. However, there needs to be intelligence involved in the process for that to occur. For example, the sun's energy is a standard source that an evolutionist uses to support how more order occurred. The problem is that if life doesn't have the necessary information already in the DNA to utilize the sun's energy; only destruction results from the sun's energy. Take the human body; we already have information in our DNA code that allows us to utilize the sun's energy to produce vitamin D; we also already have melanocytes in our dermis that have information about how to protect us from the sun's harmful ultraviolet rays and that reduce the amount of UV rays that penetrate our skin to damage our collagen and adversely affect our genomes. Without the necessary DNA code, humans would only decay from the sun's energy. Case in point, in the vinyl roof of a car, there is no intelligence to utilize the sun's energy, and the roof obeys entropy as it tends toward decay. And take botanical life forms; they have the necessary DNA code to utilize the sun's energy for photosynthesis. Without already programmed DNA code, the sun's energy is never utilized to tend toward more order. The sun's energy is only harmful and destructive.

Review: The concept of evolution, of going from a simple single-celled organism to the complexities of today, violates the second law of thermodynamics (entropy).

Even H2O already has a design inherent in its structure, which allows water to form and bond together due to the covalent bonds of H2O at a 109.5° angle. Sophisticated evolutionists use a clever argument of a snowflake to support the idea that things can tend toward order, as evolution tends toward more complexities. They will answer the violation of entropy with a question to a creationist, "Can order come from disorder?" All untrained students will say no. Then the evolutionist will show pictures of a snowflake, with the inference that order, in the form of snowflakes, arose out of the chaos of random raindrops. Then they demand that you can never say that evolution violates the second law of thermodynamics and entropy. *Photo credit: Google search: pinterest.com.*

But it's a trick question. The question should be, "Does order come from disorder without intelligence?" Then the answer is always no. Let's take a look at this snowflake trap. There is an overall loss of energy as the molecules slow down with more and more loss of energy as they cool. Thus, although the order increased in the crystallization of the snowflake, the overall system tended toward entropy with the loss of energy/heat. And the salient point is the molecules are positioned according to their designed structure—the intelligent design. Within the design of a water molecule, the position of the "H" molecule with the "O" molecule causes the location and positioning of other molecules to become attached. The snowflake indicates design because of how the molecules are positioned according to the 109° angle of covalent bonds and how they crystallize based on altitude, pressure, and temperature. The bottom line is that there is evidence of design in the water molecule, and this design is what shapes how a snowflake is formed. And to answer the evolutionists' question: yes, it appears a snowflake is more orderly, but it is based on the preexisting design order of a water molecule that dictates the order of a snowflake, and the overall system results in the loss of energy, so even a snowflake is in harmony with entropy, as are all things.

Evolutionists also argue that adding energy results in less entropy and more order and that the sun's energy is the outside source that is applied to the earth. Thus, they say that evolution tending toward more order is not in violation of the second law of thermodynamics—entropy. But adding energy doesn't add order unless a design is involved in terms of how to utilize the energy. Energy is purely destructive by itself unless there is a design that determines how to utilize the energy. Evolutionists may cite a plant as evidence of random glucose transforming into an orderly state. A plant takes CO_2 in air, water, and nutrients from the soil and the sun's energy and converts it to energy. However, with closer inspection, there is an overall lose of energy, and there is intelligence involved in terms of how to utilize the energy. The plant took a lot of energy to absorb the water and sunlight, and there was DNA in the plant telling it what to do with the energy and how to utilize the energy. Energy from the water, soil, and sun went into the plant, but the DNA was there to govern the process.

Review: The formation of a snowflake has an overall loss of energy—heat, and is based on the intelligent design inherent in the angle at which the hydrogen atom bonds to the oxygen atom. This angle and the altitude, barometric pressure, and temperature determine the finished design of a snowflake. Therefore, evolutionists cannot use the snowflake as an explanation for allowing evolution to violate entropy. Also, the DNA of a plant is intelligently designed and utilizes the sun's energy. Energy by itself without intelligence is destructive. Either way, intelligence is always required to utilize energy, and evolution still violates entropy, a law of physics.

Evolution violates another law of science. This law was first discovered by Louis Pasteur in the late 1800s. Pasteur discovered that life has not and cannot be spontaneously generated from nonliving material. He proved that life only comes from life. This is called biogenesis, which means "all life from life." Evolutionists use several methods to get around these violations of the laws of physics, as listed below.

1. Time
"Given so much time, the 'impossible' becomes possible, the possible probable, and the probable virtually certain. One has only to wait; time performs the miracles."

Evolutionists argue that given a long enough time, the impossible becomes certain. This violates simple mathematics, which says that any number multiplied by zero equals zero. Saying that "time performs the miracles" reveals that the hypothesis of evolution is a faith-based belief system. What is a miracle? It's an extraordinary event, a marvel, and a wonder manifesting divine intervention in human affairs.

For evolutionists to say that they believe in the miracle of time, which assures an extraordinary event that has never been done before or since, is leaving the world of science and unbiased opinion and entering into the realm of a faith-based belief system and myth. When they argue that they have science to back them up and that they don't believe in a fairy-tale god, they are forgetting that their belief that says time performs miracles is (a) unobservable and (b) has failed all scientific tests. Observability and testability are the two bedrocks of the scientific method.

2. Terminology

The idea is to bury any violation of the laws of physics or logic in scientific terminology that confuses the populace. For example, Darwinian evolutionists won't say they believe in spontaneous generation of the first life, but they will say they believe in abiogenesis. What? This is double talk, as abiogenesis is life that comes from nonliving material; they have the same definition. The saying "you don't get something from nothing" is appropriate. Some evolutionists have even used debating tricks, such as abiogenesis has nothing to do with evolution because evolution is only about life evolving. However, the concept of life evolving cannot occur unless chemical evolution evolves life. And according to evolution, the first life came about by abiogensis. Thus, evolution has everything to do with abiogenesis.

How about the phrase, "ontogeny recapitulates phylogeny," by Ernst Haeckel in the late 1800s? Let's break down these fancy words.

Ontogeny is the growth (size change) and development (shape change) of an individual organism from an embryo to an adult.

Recapitulate is to restate briefly or summarize.

Phylogeny is the evolutionary history of a particular group of organisms, as depicted in a family tree.

Ernst Haeckel is a well-known scientist who so dogmatically supported Darwinian evolution that he exaggerated the similarities of embryonic development of different creatures into his infamous embryonic drawings. Although his notorious drawing has been rejected by modern biologists as false information, it still exists in some textbooks for students to learn how evolution works and how one kind of creature evolves into another kind of creature.

In a debate class that I took in college, I learned quickly that the one speaking truth doesn't necessarily win a debate. Sometimes, the winner of a debate goes to the one who uses bigger words. I enjoy listening to Richard Dawkins (evolution philosopher) speak, and he would destroy most believers with his razor sharp intellect, but that doesn't mean he is on the side of truth. The reality is that sometimes the simplest answers are on the side of truth. And that's why pearls of wisdom can come out of the mouths of babes. Matthew 21:14–16 discusses "children who were shouting in the temple [to Jesus], 'Hosanna to the Son of David.'" These children were talking about Jesus the savior and rightful King of Israel. But the erudite Pharisees, in their superior piousness and prowess of intellect, became indignant at the thought of children who were so unworthy to speak about this information—let alone in the temple—and rebuked the children and shooed them away. But Jesus rebuked them back for not understanding the Scriptures by saying, "Have you never read, 'Out of the mouth of infants and nursing babies You have prepared praise for Yourself'?" Since God is King of the Jews (I Samuel 8:7) and King David wrote Psalm 8:2—in which Jesus quotes that God prepares praise for Himself out of the mouths of babes—Jesus was declaring Himself to be God in the flesh, the representative of the invisible God the Father, by accepting the praise of the children and the idea that Jesus was the Hosanna (the one to save now) and the heir to King David. Jesus was declaring that He was the Savior, King of the Jews/Israel and God in the flesh. This theme is repeated over and over in the Bible.

3. The half truth tale and the unknowing hearer

The law of entropy states that the sum of everything in a closed system will tend toward disorder. This means that some things can tend toward more order, but only when something else is applying a controlled force, with intelligence involved. However, the total sum tends towards disorder.

For example, my daughter's room tends toward disorder; in order for the room to become organized, a controlled intelligent force needs to be applied. She can't just throw a firecracker in the room and expect a "big bang" to organize everything. She also can't just start a fire in the room and apply unintelligent energy. My daughter needs to expend energy, with intelligence, to put order into her room. This expenditure of energy comes at a cost; she needs to eat to gain the fuel to have energy to clean her room. After she has cleaned her room, she has spent more energy cleaning than is actually applied to the actions of cleaning and organizing. Therefore, the overall net sum tends toward disorder, even though her room has been cleaned.

Evolutionists argue that life evolves toward more order and more complex kinds of creatures because another outside system is applying force, which causes the reversal of entropy. The outside system is either the sun, which provides energy for life on earth, or other universes that have applied a force to create order within our universe. Evolutionists argue that the earth is not a closed system.

Is it true that the sun applies energy to cause life to move toward more complexity and more order? If the sun's energy didn't have a powerful destructive component, then the answer would be yes. But the sun's energy is destructive as well as constructive. Imagine if my daughter tried cleaning her room by removing the roof and letting the sun's energy clean her room. Since the sun is not an intelligent force applying energy in exact locations and in guided directions, then more often than not, the sun's energy that is applied to her room would cause more entropy. And similarly, the sun's energy to the earth accelerates disorder more than it advances order. And every time there are advances in order, intelligence is always involved. Therefore, this outside source utilized by evolutionists to say evolution doesn't violate the law of entropy is merely a well-constructed argument based on half truth. The unknowing hearer is unable to argue back.

Don't forget about the sun's ability to send destructive solar flares that are harmful to life on earth and that increase entropy tremendously. Also, ultraviolet light that accelerates the aging process, which is code for increasing disorder, increases entropy. Also, it should be pointed out that there is a whole lot of entropy in the Sahara Desert and many other deserts across the globe that results from the sun's unintelligent energy. These deserts cause very hot wind to move, and when that heated air comes across moisture, then severe destructive weather patterns can occur. A whole lot of entropy is created in the aftermath of tornadoes and hurricanes.

In addition to the sun's energy being unintelligent and it's energy being more destructive than constructive, there are other outside sources on the earth, such as the moon. The moon is receding from the earth at 3 cm per year. The moon is taking angular momentum from the earth to move away from the earth; thus, the spin of the earth is slowing slightly each century and the moon is aiding in the earth moving toward disorder. The sun applies unintelligent energy to the earth, the moon steals energy from the earth, and the cold of space takes energy from the earth via heat transfer. Then there are other outside sources that apply harmful energy to the earth, such as gamma rays and cosmic rays from exploding stars. Other outside systems, such as asteroids and comets, bring their own destructive energy to the earth to increase disorder and to increase entropy. Outside systems bring more entropy upon the earth than those that bring energy that is usable for increased order and the complexities of life. And without intelligence to utilize the sun's energy, there would be zero increased order. The sun is slowly dying, losing energy. One day in the distant future (billions of years from now), the sun will no longer provide light, heat, or energy. This means that the sun is tending toward decay, increasing entropy. Thus, even though the sun provides energy to the earth, and intelligence applies that energy toward more order, the overall sum of entropy increases.

Even though those prior arguments are sound, there is a more obvious area to explore. Since the sun is an unintelligent source of energy, it doesn't apply its energy at one thing in a controlled manner. If the earth is becoming more orderly and more complex from the sun's energy just like ALL life is allegedly evolving into more and more order and complexities, then the evolutionists have an argument. But the earth is gradually moving more and more toward disorder and is in harmony with entropy. This is seen with the slowing of the earth's spin, with the buildup of past living organisms adding to soil content, the decline in the magnetism of the poles, the decline in the protective ozone layer, shorelines eroded by the ocean, earthquakes, tornadoes, hurricanes, and so on.

Therefore, since the earth as a whole is moving toward more disorder and is increasing in entropy and since the sun's energy is unguided, unintelligent, and impartial, then one can't argue that life on earth is using the sun's energy to generate more order and more complexities without intelligent design to utilize the energy. Thus, the primordial soup of complex chemicals that lacked intelligence in the form of DNA, could not utilize the sun's energy to increase order into the first replicating single-celled organism. If the earth was becoming more orderly and moving toward a more complex system, then life on earth would too. But since the earth is moving toward disorder and a less complex system, then life on earth would also move in the same direction if intelligent design wasn't behind the existence of life.

Review: The sun's energy is unguided, unintelligent, and impartial and is more destructive than constructive; it is an outside source assisting the planet to move toward more disorder and more entropy, not less entropy and more complexity. Therefore, one cannot count on the sun to save evolution from its violation of the second law of thermodynamics—entropy. Evolution violates physics and tries to cover up this violation with half truths.

4. Argumentum ad hominem

Argumentum ad hominem is a Latin phrase that means arguing against the person versus the ideas of the person. In this case, it applies to those who make personal attacks against nonbelievers of the hypothesis of evolution versus arguing against the evidence against evolution. Personal attacks include being called a fool, idiot, and so on. This stifles a reasonable, logical debate.

Since many believers in the Darwinian evolutionary theory are atheists, they are not subject to absolute morality from a higher authority that commands obedience to the concepts of kindness, gentleness, self-control, humility, patience, peace, joy, and love. Therefore, the personal attacks can be quite severe, with nothing to restrain the verbal assaults. Some Christians have lost their jobs for voicing their concern about the theory of evolution. And some Christians have lost their freedom for exposing evolution as a lie (for example, Dr. Kent Hovind). Although Dr. Hovind was in jail for 10+ years, he still has a powerful ministry exposing the lies in evolution.

Of course, this is a two-edge sword. There have been ample wolves in sheep's clothing that claim to represent God that do all kinds of atrocities on humanity. Of the two types of offenders, the one who proclaims to be a believer in God, yet is not and commits evil, is the worst kind.

Review: Evolutionists cover up the fact that their hypothesis violates the laws of physics by saying that time saves the day and by using Latin terminology, half truths, and argumentum ad hominem (personal attacks).

Group Discussion:

1. Now that you know that the principles of evolution violates physics, and that the Bible is in harmony with physics, how did that influence you?

2. How has gaining the knowledge of the truth that life always comes from life effected your faith in the Bible?

Chapter 30
Evolution Versus Science

Science means having knowledge in an area or knowledge that is an object of study, covering general truths or the operation of general laws, especially as obtained and tested through the scientific method.

A valid question from a scientific point of view is where did God come from? The Bible records God's existence is infinite/eternal. Does this violate science? No, in science, it is acceptable to have something be infinite. For example, the space that the universe is expanding in is said to be infinite and immeasurable. In mathematics, the numbers to the right of the decimal point continue on forever, and there is even an infinity symbol: ∞. In astronomy, the edge of the universe that is expanding is boundless. In geometry, a line is infinite. Most things are finite, meaning they don't go on forever, so it's extremely rare to find something that is infinite. Of the innumerable things in the universe, there are only a couple of infinite things. Evolutionists hypothesize that energy and some laws of physics have existed forever, but God could not? That is a personal bias.

Does evolution violate science with the belief that life and the DNA information within all life spontaneously generated from nonliving material? In 1958, George Wald wrote,

> Throughout our history we have entertained two kinds of views of the origin of life: one that life was created supernaturally, the other that it arose "spontaneously" from nonliving material . . . This great controversy ended in the mid-19th century with the experiments of Louis Pasteur, which seemed to dispose finally of the possibility of spontaneous generation. For almost a century afterward biologists proudly taught their students this history and the firm conclusion that spontaneous generation had been scientifically refuted and could not possibly occur . . . <u>Conceding that spontaneous generation does not occur on earth under present circumstances, it asks how, under circumstances that prevailed earlier upon this planet, spontaneous generation did occur</u> and was the source of the earliest living organisms . . . Given so much time, the "impossible" becomes possible, the possible probable, and the probable virtually certain. One has only to wait; time performs the miracles.

Is there a violation of science with the evolutionary belief in abiogenesis, the spontaneous generation of life from nonliving material? Let's dive into the science and determine whether evolution or the Bible violates science. Scientists have created a basic template for which a hypothesis (an idea to explain a phenomenon) is measured to determine if the hypothesis is worthy to be considered a scientific theory (an idea accepted by the scientific community).

The scientific method has specific steps of solving scientific problems:
    * Make an observation.
    * Ask a question.
    * Form a hypothesis.
    * Conduct an experiment/test.
    * Analyze the data.
    * Either accept or reject the hypothesis.

Evolutionists claim to be on the side of science, but the very first step of the scientific method is to "make an observation." George Wald, speaking on behalf of evolution, states, "<u>Conceding that spontaneous generation does not occur on earth under present circumstances, it asks how, under circumstances that prevailed earlier upon this planet, spontaneous generation did occur.</u>" There has never been an observation of spontaneous generation of life from nonliving material, and all the experiments to support the idea of spontaneous generation from nonliving material have failed. The two most important steps of the scientific method, observing and testing, have failed. Therefore, the hypothesis of spontaneous generation must be rejected. But instead of rejecting the hypothesis, it's taught as fact. The assumption that "conditions must have prevailed upon the earth to generate spontaneous life from nonliving material given enough time" is not science; it is a leap of faith.

There are two leaps of faith that an evolutionist must take: (1) since spontaneous life doesn't occur today, then the conditions that prevailed upon the earth to generate spontaneous life from nonliving material were significantly different than today, and (2) this difference allowed abiogenesis to violate the laws of biology because of the length of time that performed the miracle of life. One of the differences they assume is that there was no oxygen on the planet at the time of this spontaneous generation of life. This is not observable, and to compound the error, evolutionists assume that conditions were perfect to have spontaneously generated life that wouldn't have spontaneously generated in earth's current conditions. And building upon those two leaps of faith, spontaneous generation was certain to occur given enough time. The time aspect is testable, but unfortunately, people interpret data and set up tests to support their preexisting beliefs in regard to an old earth, rather than actually seeking truth. Remember that the flawed testing of time is based on a constant rate of decay (CRD) that is not constant, and this is how they determine the earth to be very old.

This next quote from, *Principles of Biochemistry* (Lehninger, Ch. 3, pg. 59–62, sixth printing 1988), a college biochemistry textbook illustrates the nomenclature of speculation that is used to lay a foundation on which to build facts.

> It has been suggested that all living organisms may have descended from a single primordial cell line. Thus the first cells to have arisen on earth and survived may have been built from only a few dozen different organic molecules which happened to have, singly and collectively, the most appropriate combination of chemicals and physical properties for carrying out the basic energy transforming and self-replicating features of a living cell . . . But here we have a dilemma. Apart from their occurrence in living organisms, organic compounds, including basic biomolecules, occur only in traces in the earth's crust today.

This biochemistry textbook is espousing the belief that life spontaneously generated from nonliving material, acknowledging that the nonliving material exists in trace amounts today. Where did all the nonliving material go? Where did all the primordial complex chemical soups go that allegedly formed life? It's fine to have a belief, but do not call it science. The text book goes on to say,

> It is believed that the earliest living cells eventually used up the organic compounds of the seas, not only as building blocks for their own structures but also as nutrients or fuel, to provide themselves with the energy required for growth. Gradually, through the ages, the organic compounds of the primitive sea were consumed, faster than they were created by natural forces. As organic molecules disappeared from the seas, living organisms began to "learn" how to make their own organic biomolecules. They learned to use the energy of sunlight through photosynthesis to make sugars and other organic molecules from carbon dioxide; they learned to fix atmospheric nitrogen and convert it into nitrogenous biomolecules, such as amino acids.

We cannot see the primordial chemical soups because the initial organisms used up nearly all the organic compounds? How convenient, and the proof that abundant amounts of primordial chemical soups once existed is that they are in trace amounts today? And the reason they are in trace amounts is proof that the organisms ate them? This is mythological, not science. This is a classic ploy of reverse logic, that because the primordial soup is in trace amounts and is not natural to Earth, it proves that it got consumed and initially was abundant. That is an illogical assumption. The mistake is believing that a few organic elements, such as nitrogen, carbon, hydrogen, and oxygen, could form together to make a functioning cell with programmed information in the form of DNA. This is the equivalent of dirt that has all the compounds of a car; given enough time, the car emerges under the right conditions and consumes all the remaining leftover compounds as fuel and then learns to consume a different fuel source. This is beyond the realm of possibility. Even debating where the information came from for the living organisms to learn to consume a new fuel source and to learn how to utilize the sun's energy via photosynthesis is going beyond the threshold of the possible—let alone discussing spontaneous generation of life. The huge leaps of faith include the following.

A. Organic molecules occur in trace amounts naturally, so to presume that there was an abundance at one time for life to spontaneously spawn is faith based.

B. To cover up the loophole by espousing that the reason organic molecules occur in trace amounts is because the early life forms used it all up is faith based.

C. The idea that after using up all the organic molecules, life evolved the ability to find and consume a new source of fuel is faith based.

Then after these faith-based assumptions, evolutionists proclaim they are on the side of science and do not require faith.

Review: The scientific method requires observations and tests to determine if a hypothesis can be a theory. Evolution relies on faith-based assumptions and then builds on that.

Evolutionists proclaim that there is a natural process that does all the work that God is supposed to do or has done. And that natural process is via natural selection (survival of the fittest) and unguided, random mutations of DNA. These two natural processes cause all life forms to become more complex and diverse. There are several problems with this hypothesis; one is that natural selection gets rid of the mutated and frail and weak to preserve the better genetic copy of the original. And all mutations produce a loss of a function that leads to frailty, weakness, sterility, stillbirth, and so forth because all mutations result in a loss of information in the DNA code. Both natural selection and random, unguided mutations of the genetic code actually prevent changes in the genetic code from being passed on. Both ensure that the closest exact copy of the original genetic code is preserved by getting rid of creatures that have errors in their genetic code. In reality, these two things, natural selection and genetic mutations, function in reverse of what evolutionists contend. Natural selection and genetic mutations both preserve God's creation; they preserve the purity of the genetic code as best as possible because those with mutations in the genetic code are eliminated via the natural selection process. These concepts that form the very foundation of evolution actually prevent evolution. Isn't it just like Satan to take what God has created (natural selection) and twist it for his benefit?

By the way, it should be noted that natural selection—which allows a predator to seek out the weak and kill it amongst a herd of the strong, and implants the desire of the female to only mate with the best genetic copy of the original—is already built into the DNA. There is no new information; in fact, the existing DNA information is designed by God to prevent new information that could result in a new kind or a new function. A beak that adapts to grab food better is based on DNA that already exists in the code. When I talk about a new function, I'm talking about (for example) a bird evolving baleen teeth or some photosynthetic skin cells, which would mean that the bird would and no longer need to eat with a beak.

The ability of a life form to adapt to its environment is accepted as fact by creationists, but there are limits to how far adaptation can go. When the edge of adaptation is reached, the limitations are set with stillbirths, sterility, and deformities, and these limitations are determined by the DNA in terms of how far a particular kind of creature can adapt to external conditions. Adaptation is observable and accepted as fact, but the creature adapting always remains the same kind of creature. Never ever has there been an observation of a kind of creature adapting so far that another kind of creature is formed. God has established in the DNA that each kind of creature remains the same kind.

Review: Random, unguided mutations in the DNA code always result in a loss of function or an impairment of function, and then natural selection eliminates those mutated creatures to preserve the DNA. All creatures with mutations are either weak, frail, sterile, stillborn, and so on, and natural selection eliminates them. God has determined the boundaries of adaptation, and that boundary is hardwired in the DNA code. Also, natural selection is one of the means that God uses to rid the world of mutations. Both the preexisting DNA code and natural selection prevent evolution from occurring.

There is an ongoing experiment to determine how Escherichia coli bacteria adapt to different stimuli and to ascertain how/what bacteria can evolve into. This experiment is frequently cited by evolutionists as proof that evolution does indeed occur. The end result after 50,000 generations (which is more than the entire history of all humans and primates combined) of E. coli reproducing in every conceivable environment is still E. coli bacteria, with different adaptive characteristic. This is not evolution; this is adaptation. Now, if the test started with E. coli bacteria and finished with a virus, blade of grass, a worm, or a fungus, then evolutionists would have something. But after 50,000 generations, bacteria are still bacteria, which establishes that all kinds remain the same kind. Living things have the ability to adapt to changes in the environment, yet there are limits in terms of how far life can adapt and survive.

Science is always welcomed by the Bible. No creationist fears chemistry, physics, or biology, as they are obedient servants to God. One will never find the Bible in violation of any laws or principles of science. There is bilateral harmony because the God that fine-tuned the universe, created life within it, and recorded His actions by the pen of mankind through the inspiration of the Holy Spirit is God of all and creator of all, and that includes the laws of physics. He is not just the creator of man and life, but also of all the laws of physics, biology, and chemistry. Don't confuse man's discovery of these laws with the origin of these laws; they were there before man had the ability to understand modern science and to discover these laws; that's why—as man progresses with science—we still find hidden truths in the Bible that are in harmony with the sciences.

I was debating an evolutionist about, and he argued that theists believe that only intelligent design can breed sheep together to produce woolier animals through breeding (artificial selection), but a series of very cold winters would also create woolier sheep as well, via survival of the fittest (natural selection) and random mutations that alter the genetic code. But this is not what intelligent design proponents are arguing, as they are not saying that woolier breeds can only be created through breeding. Intelligent design proponents are saying that the information necessary to produce a woolier sheep was already coded in the DNA way before any breeder came along and way before any winter came along and caused natural selection. The information was in the sheep's DNA at birth, from its parents, and their parents got their DNA from their parents, and so forth, until you get to creation. It's not that cold winters over several thousand years produces new DNA code to produce woolier sheep. That's not how it works. Similarly, the potential for new species within the same kind of creature via adaptation, whether by natural selection or artificial selection, is also already coded in the DNA. Therefore, arguing that natural selection rules out any God is a poor argument. Creationists accept as scientific fact that adaptation occurs, and given the same parameters of cold winters, the sheep with a woollier coat would survive, and when they reproduce they would pass on a genome with information for a woolier coat already embedded in the DNA from their ancestors, not from a new DNA code produced because of cold weather.

Dr. Stephen Meyer is on the cutting edge of using the scientific method for authenticating intelligent design. His book *Signature in the Cell* is a must read for anyone willing to fully understand the improbability of evolution and the conclusive deductive reasoning behind intelligent design. Dr. Meyer says that all information, whether it be from newspaper, books, computer program, or in binary form is exclusively acquired through intelligence, especially in regard to information that has a function. This is exactly what DNA consists of: digital and functional information. Therefore, based on all the experiences we've had, whenever we see information, especially functional digital information, we always assume intelligence, and we are always correct.

Review: Sheep adapt to cold and become woolier because the information is already hardwired into the genetic code. There is no random, unguided mutation of the DNA code that causes sheep to be woolier. Natural and artificial selection do NOT create new information in the DNA code. There is a difference between adaptation and evolution; adaptation already has the DNA information to adapt, and evolution allegedly makes new code via random, unguided DNA changes that produce a new function or a new "kind."

Evolutionists argue that the proof that man evolved from chimpanzees is that we have an almost identical genetic code, and there is only a 1% difference in the DNA between human beings and chimpanzee. We aren't talking about lay people saying this; there are professors that sneak this information out to unprepared students and into debates. When believers in God hear this, they are left thinking, *well then, maybe they are correct*, and students who don't know God, are all the more convinced of the viability of evolution. How do evolutionists justify that we share 99% similarities in the DNA? By teaching that when there is a dissimilarity in the DNA between chimps and humans, they discount this DNA as non-functional junk DNA. However, according to geneticists Dr. John Sanford, there are no non-functional junk DNA strands in the code, and thus, we only share ~50% similarities in the DNA with chimps. Notwithstanding, even if we falsely assume there are non-functional junk strands of DNA, Dr. Grady McMurtry explains that that evolutionists are not telling the whole story because the claim that we share 99% DNA is based on comparing only 2.7% of the DNA code from humans and chimps. Yes, the human DNA code has been 100% mapped out, but not the chimps'. Now, why would one make a bold claim that we evolved from chimps, while leaving out 97.3% of the information? According to Dr. Grady McMurtry, chimps and humans do not share 99% as evolutionists claim. Only 83% of the DNA is similar. But 88% of the DNA shared by humans and rats is similar, and humans share 60% with bananas, and sea squirts (marine invertebrate creatures living near coral that look like tubular flowers) share 88%. The reason there are similarities of DNA strands is because we literally came from the same creator—from the same essence through God. And secondly, there are processes within living things that share similar necessities and functions, and those similar properties or components result in similarities in the DNA code.

For example, all creatures that breathe oxygen and have blood running through their bodies have similarities in the DNA code because they share similar functions. Likewise, all life forms that have similar functions have a similar DNA strand. This doesn't mean that we evolved; it means that we came from the same creator. When evolutionists argue this point, they equally argue for God without knowing it.

Review: The argument that we share 99% of chimps' DNA is false. The empirical data is that we share only ~50% similar DNA, and not just with chimps, but with many creatures. Creatures that have similar characteristics have some DNA similarities by default. This illustrates that we didn't evolved, but that we came from the same creator who was efficient with using the same DNA code that produced similar functions in many life forms.

In life, when something appears obvious, don't overthink it. Richard Dawkins said, "Biology is the study of complicated things that give the appearance of having been designed for a purpose."

The question is, is there an appearance of design, or is there really design in life? How did the concept of intelligent design get downgraded by those in the late 19th and early 20th centuries, and why was it removed from "science"? Well the forefathers of evolution in the 19th and early 20th centuries formulated theories on spontaneous generation without full knowledge of cells, amino acids, proteins, DNA, nucleotides, and so on. For example, the idea that a single-cell organism could spontaneously generate from nonliving material was spawned by a misunderstanding of the complexity of cells. They didn't understand cells at all. They thought the primordial pool of complex chemicals simply had to crystallize for life to form the first single-cell living organism. For example, in 1925, Ernst Haeckel said, the cell is a simple "homogenous globule of plasma," a gooey ooze of the same structureless substance. This type of erroneous understanding was the basis for thinking that life could evolve from nonliving material. Today, evolutionists still cover up this error concerning the spontaneous generation of life; they still try to prove that their belief is still a valid hypothesis, even though it has never been observed and has failed every test. But they opted to force their hypothesis by skipping the scientific method and going straight to saying, "It's fact." Anyone who says otherwise is going against science.

Watson and Crick discovered DNA, and now we know that DNA has digital-like characters arranged according to a preexisting code that tells the building blocks how to form a protein and where to go and what to do. That preexisting code came from the initial fertilization of parental DNA. Therefore, we have to keep going backward in time to find the origin of the information. To the evolutionist, the first single-cell prokaryote got its information from nonliving material via the spontaneous generation of information. To the creationist, the first man got his information from an all-knowing God (life begot life). God infused Adam with parts of God's knowledge. Evolutionists say that life was spontaneously generating from nonliving material, suggesting that genetic information was also spontaneously generating from nonliving material. Without the information in the DNA code, there would be no life. There would be no knowledge about how to find food, consume food, process food, mate/replicate/reproduce, defend oneself, get rid of waste, walk, and so on. This is a huge problem for evolutionists, but they just shrug it off and proclaim that the information evolved through random mutations over time. This is faith based.

Review: The forefathers of evolution perceived cells to be chemically and biologically simple, which made the process of going from simple amino acids to a living cell simply through crystallization seemingly plausible. But reality couldn't be further from that error. Information from DNA is vital for life; without information to know how to innately function on a macro- and micro-cellular basis, life would not exist. The spontaneous formation of DNA information for the first living thing is another faith-based leap for evolutionists.

When the pioneers of evolution postulated that life spontaneously generated from nonliving material, the hypothesis was formed without full knowledge of science, and so there should be some gentleness for their ignorance. But modern-day evolutionists carry on this idea, even though there is ample evidence of the errors in the hypothesis of evolution and more knowledge about the problems with the origins and evolution of life that is readily available to evolutionists. Belief in evolution is a willful choice to accept half truths and huge leaps of faith, instead of seeking truth.

There were several complexities that Ernst Haeckel and Darwin and the early evolutionary forefathers were unaware of regarding cells. For example, there are four elements that make up an amino acid: oxygen, hydrogen, nitrogen, and carbon. Also, there are 20 amino acids that are common in proteins: glycine, alanine, valine, leucine, isoleucine, methionine, phenylalanine, tryptophan, proline, serine, threonine, cysteine, tyrosine, asparagine, glutamine, aspartic acid, glutamic acid, lysine, arginine, and histidine.

The sequence of the above 20 amino acids that form a protein are in a precise order of about 150 amino acids long; this order is based on a predetermined, preprogrammed DNA to form a single protein. One amino acid out of place results in malformation, which adversely affects the function built upon that protein. The DNA information is stored in digital form, which means it's similar to a computer program that determines how the amino acids are formed to create the three-dimensional shape of proteins. One error in the sequencing for a piece of protein formation alters its shape, and it won't fit into a receptor sight to carry out its function. Preexisting DNA information from a prior life determines the sequencing of the amino acids that form proteins. Proteins are vital for functions within the cell and the entire body. Since there are 20 amino acids, and the chain of amino acids to form a protein is 150 long, and since evolutionists believe that random chance evolved the first protein and not DNA, then what are the odds to produce one single protein? The calculation is 1 in 20 x 1 in 20 x 1 in 20, for 150 times. The odds are ~1 in 2×10^{195} for random, unguided chance to form one single protein (this is logically impossible). To put that in perspective, the average six-number lottery with 1–49 numbers, the odds are one in 1.4×10^7 to win the lottery. And this doesn't include the more complex process of forming DNA structures. Which one do you have more faith in, winning the lottery or winning the evolutionary hypothesis lottery to form one protein?

The odds to form one protein through random, unguided mutations is astronomically low; we don't need one successful event for the first life; we need this random, unguided chance to be successful 250–500 times for each cell because each cell on average has that many proteins. The odds get worse; each protein has around 1,500 genes in proper sequence.

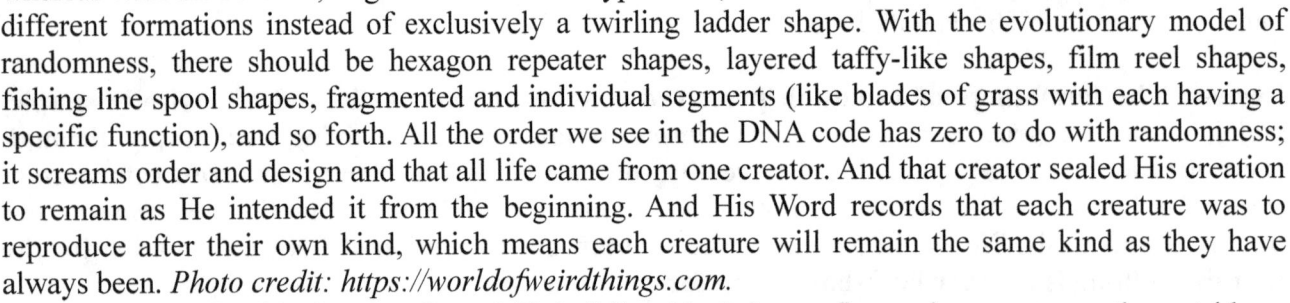

There are four nucleotides in DNA that form genetic code: adenine, guanine, cytosine, and thymine. This is where the A, G, C, and T letters come from when one sees a DNA helical structure. It is important to note that the double helical structure of DNA, from what we have studied so far, is always in the clockwise formation. This represents design and order, whereas with the random, unguided mutations hypothesis, there should be different formations instead of exclusively a twirling ladder shape. With the evolutionary model of randomness, there should be hexagon repeater shapes, layered taffy-like shapes, film reel shapes, fishing line spool shapes, fragmented and individual segments (like blades of grass with each having a specific function), and so forth. All the order we see in the DNA code has zero to do with randomness; it screams order and design and that all life came from one creator. And that creator sealed His creation to remain as He intended it from the beginning. And His Word records that each creature was to reproduce after their own kind, which means each creature will remain the same kind as they have always been. *Photo credit: https://worldofweirdthings.com.*

Each nucleotide (A, C , G, and T) building block has a five-carbon sugar on the outside, a nitrogenous base, and a phosphate group. They make up the DNA code, which determines the precise order of the above 20 amino acids that form the 150-long chain needed to make one single protein. With random, unguided mutations forming the DNA coding, there would be errors in the DNA. And these errors would adversely affect the sequencing information that instructs the proper order of the 150-long chain (consisting of the 20 amino acids) that are needed to form one single protein. Without a precise sequencing of the amino acids to form the protein, then there is a complete loss of structure and function of the protein. The protein is a precise, complex 3-D shape that fits other molecules like a hand in a glove and that carries out exact functions—one amino acid in the wrong order would cause a total failure of the protein because there wouldn't be a perfect fit with the other molecule that it's designed for to carry out the exact function. Therefore, nothing occurs—no function, no catalyzed reactions, no building structural parts, and no processing of DNA. This is just like a computer program; make one coding change randomly, and the entire function is broken. And where does the information come from to put together the exact order of the amino acids to form one single protein? The DNA code has the information for which to build the different proteins that perform different functions for the cells.

There are four nucleotides (A, G, C, and T) options for a sequence 1,500-chain long. Therefore, the odds of random, unguided chance forming the proper gene for one solitary protein is ~1 in 2×10^{903}. Plus, add the ±1,500 properly sequenced genes for each of those 250–500 different proteins, and you quickly start to see the problem. There is a lot of information required. The odds of random chance forming the necessary information to form one cell is beyond possible. The odds of random, unguided mutations forming the DNA code, which has vastly more information, is beyond impossible. This is why chance via random, unguided mutations in the DNA code would most likely degrade a code rather than enhance it. Think about it; the DNA code is where the information comes from to sequence the 20 amino acids in proper order 150 times to form one single protein, so without the DNA, there is no information to guide the sequencing order. This is where faith in evolution takes over science.

Additionally, if an evolutionist argues that the nucleotides (A, G, C, and T) are attracted to each other because of charge, and there is no need to rely on the low odds of random chance to form the DNA sequence, this is not accurate. For the sugars and phosphate that form the backbone (or structure) of the helical DNA structure are chemically bonded by electrical charge, but they carry no information, zero. And there is a chemical bond between each nucleotide to the sugar-phosphate back of DNA, but there is no bond and no attraction among the nucleotides (A, G, C, and T). The sugar-phosphate bond allows any one of the four nucleotides to bond with it, but there is no bond or attraction between each of the letters of the nucleotides. Therefore, one cannot argue that the DNA sequence of the nucleotides is based on attraction. There is no attraction between the nucleotides that result in the DNA sequence holding information (Stephen Meyer, *Signature in the Cell*). This means that evolution is stuck with random, unguided chance to form the first DNA sequence and to form new information for new functions and for new kinds of creatures. This is an untenable hypothesis.

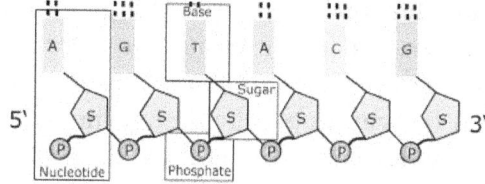

Image adapted from: National Human Genome Research Institute.

But it gets worse for evolutionists, as we haven't even discussed the odds of a select few elements, randomly with no guidance, forming together to form an amino acid to begin with. They say that nitrogen, oxygen, carbon, and hydrogen randomly came together from the first rains on the rocks on the earth some four billion years ago. Therefore, if the select elements (N, O, C, and H) formed together by random unguided chance to form the 20 amino acids, then the odds get worse factoring in the formation of the 20 amino acids. Then we have to calculate the odds of random chance forming the elements in the first place. The odds are never in the evolutionist's favor.

How does an evolutionist sidestep this problem? They either say that given enough time and enough multiple universes, life is bound to begin in one of them and form new functions and new life through random, unguided mutations. They just bypass the reality of the impossible odds and just regurgitate the evolutionary doctrine, like a chanting monk. Another more astute evolutionist would agree that chance by itself is not sufficient, but what can't be accomplished via chance could be accomplished through natural selection. The problem with this is that for natural selection to work, there must be sexual or asexual reproduction. For reproduction to occur, there needs to be a specific DNA sequence with specific proteins to carry out reproduction. Therefore, the information is required first. The information is what causes reproduction and natural selection. Natural selection does not provide the information, or cause new information, it can prevent bad information. But believing that natural selection helps shape or form new information in the DNA code is putting the proverbial cart before the horse.

As discussed earlier, there is no bond between each nucleotide to other nucleotides. Therefore, there is no law of attraction that explains how the sequence of the DNA is formed. And for this reason, the theory that says the law of attraction, given enough time, will form the DNA sequence is completely false. This poses a question: "Where did the information come from to form the DNA sequence in the exact proper order of 1,500 sequential chain of nucleotides to form one single protein, multiplied by 250–500 times to form the number of proteins in one cell?" Always, this information comes from a prior life, and that is where God comes in. God is the one who provided the knowledge, but the evolutionist are left with faith in an erroneous view that suggests that chance and time provide the miracle. The Bible declares that there is no such thing as chance and that even the outcome of the lot that is cast (like dice) is of the Lord (Proverbs 16:33).

Review: The odds of winning the lottery: 1 in 1.4 x 10^7. The odds of evolution winning the protein formation lottery: ~1 in 2 x 10^{195}. The odds of evolution winning the genes that form one protein lottery: ~1 in 2 x 10^{903}. The complexities of 20 amino acids forming a long chain with 150 amino acids to make one protein is based on information predetermined by DNA coding. The complexity of generating DNA code is beyond comprehension as compared to creating a single protein. Putting this information together reveals the impossibility of random, unguided mutations forming a single amino acid, let alone a chain of 150 of them to form a protein, let alone for nucleotides to form genes for each protein and the even more complex DNA code, which is where the information comes from to form the sequencing of amino acids that form proteins in the first place. This illustrates the faith-based belief required for evolution's abiogenesis of life and information. The design and order of the DNA code suggests intelligence designed it. With the evolutionary hypothesis, there should be a random display of shapes in the DNA structure.

The cell that Ernst Haeckel erroneously hypothesized to prove evolution is not a gooey blob of plasma at all; it has complex mechanisms. An amoeba cell—which may be similar to what evolutionists believe was spontaneously generated out of nonliving material as the first prokaryote cell—moves by pseudopodia and has a genome the size of 290 billion base pairs, or DNA units, which is 100 times larger than what a human has. Though the amoeba genome is very complex, it doesn't have the number of genes that a human has. *Image Credit: https://en.wikipedia.org/wiki/Amoeba.*

An amoeba is complex. A cell must have a cell wall and a membrane around each vital organ within the cell, that is semipermeable, with the ability to engulf desirable objects for consumption (endocytosis and phagocytosis) and expel waste (exocytosis). And it must know the difference between harmful and beneficial objects. Cellular life forms have to have a means of locomotion to pursue food and flee predators, and know the difference. A cell must have a power source (mitochondria for animals and chloroplasts for plants) to provide energy for all functions. A cell must have a place to bring food into the cell (vacuole), a processing plant to convert food into energy, and a place to store processed food for future needs. A cell must have a waste-processing plant (vacuole) to expel used waste; otherwise, the cell would become toxic from waste buildup. A cell must have a central information control center (nucleus) to provide information for all the necessary functions. A cell must have a means of reproduction (nucleus), such as splitting equally into two via cellular mitosis, with both cells containing the same DNA information. Without any of the above functions, the cell will die. Without the required, preexisting DNA coding for the organ to exist and function, the cell dies. All of these processes and information are necessary for the operation of each function. And all these functions are because of preexisting DNA information. Furthermore, and I cannot stress this enough, all the above components of a cell are interdependent. If a cell is missing one components then there is no life. Therefore, all proteins, all components of a cell, and all the DNA had to be simultaneously generated for the first evolved life form.

The error of the forefathers of evolution to think the simple single-celled amoeba was a simple homogenous ooze that merely had to crystallize from a primordial soup of nonliving material could not be further from the truth. The evolutionary model suggests that not only did the first living organism spontaneously generate from nonliving material, but so too did the information that is required first before the single-celled organism can function or be alive. Therefore, one can see that evolution is a leap of faith, not science.

Review: The odds that the information needed to form the complex functions within a one-cell organism from nonliving material are too low to consider a possibility. Life and the DNA for life, always come from a prior life.

When one takes a bird's-eye view or a worm's-eye view, life always comes from a prior life, and the DNA information for life always comes from the prior life. Thus, the only allowable conclusions based on what is observed and tested, is that life and the DNA information for life can only come from a prior life. From this, there are only two options; either this pattern repeats for infinity back in time, which we know cannot be true since the earth is finite, or at one point in time an infinite life with it's DNA information begot life with DNA information on Earth.

Every attempt to explain the origin of information through evolution by chance, necessity, and time still comes down to information spontaneously generated from nonliving material. This notion is faith based. And it is not rational based on all observation and tests that affirm that life only comes from a prior life, and the DNA for life only comes from a prior life's DNA. But abiogenesis is obedient to the methodology that there is no God to start life and provide the information necessary for life.

How do evolutionists solve the dilemma of where DNA comes from? Their answer is that mRNA did it. It plays a vital role in gene expression and in coding, decoding, regulating, and catalyzing biological reactions, such as protein synthesis, by directing the assembly of proteins and by transferring amino acids to the ribosomes for manufacturing amino acids together to form proteins. However, there is a huge gap between the formation of a protein and DNA formation. The salient point is that there needs to be DNA information to instruct mRNA. Therefore, where did the information in the RNA come from to send the amino acids to the ribosomes to manufacture proteins? Another question is where did the ribosomes come from. We are still back at the same starting point, which is that information can't be spontaneously generated. This is impossible and has been proven wrong, but it's the beginning point of evolution, and all theories built upon this error are also in error.

Review: The information required for life is too complex for random, unguided sequencing of amino acids in a primordial pool to form a basic entity, such as a protein, let alone to account for the complexities of the DNA code in the simplest single-cell prokaryote life form.

All the parameters and quantitative values of the universe, such as those relevant to the laws of thermodynamics, motion, electricity, electromagnetism, light, energy, equilibrium, solids and fluids and gases, sound, heat, and so forth, require precision and finely tuned constants. If subtle changes were made to the speed of light, the laws of attraction, gravitational force, the elements of the periodic table, the constant nucleus force required to hold onto electrons, the ratio of fundamental forces, the expansion rate of the universe, or covalent bonds, then life would be impossible. This indicates that a superintelligent being set the parameters to make an environment in which life could be sustainable. Since there is so much fine-tuning of the universe and in all the properties of physics, chemistry, biology, and mathematics, then the logical conclusion is that life doesn't happen by random chance, but by the hand of an intelligent being—God.

The simple reality that our DNA is held together by hydrogen bonds is indicative of fine-tuning. If DNA were held together by covalent bonds or ionic bonds, then this tighter bond could hinder the reproduction via cellular mitosis. Hydrogen bonds are weak enough to allow their helical structure to split apart for reproduction, but covalent bonds are very strong and can keep diamonds together. Without chromosomes being split into two equal pairs, cellular multiplication for conceptional life development would be hindered. This is another example of fine-tuning within the universe but on a molecular level.

When evolutionists realize that the spontaneous generation of life and information are impossible, then they try to solve this problem with multiple universes. Thus, they speculate that a multiple-universes theory provides an opportunity where a single universe doesn't. Why can't a multiverse theory work? For one thing, what and where is the universal generating mechanism or factory? In other words, where did all the universes come from? String theory is their answer, but it's a speculative hypothesis and possibly explains how the universe became fine-tuned with all the laws and constants of physics, but not the initial conditions. Evolutionists have also introduced "universal inflation," which is another theory that could possibly explain how the universe expanded so fast in a short amount of time, but not the fine-tuning of the laws and constants of physics.

Therefore, they utilize speculative postulates to explain away what the Bible records as the work of an all-knowing creator, who set in motion all the initial conditions of a fine-tuned universe and is currently sustaining all that He created (Colossians 1:17). Both the string theory and inflation theory require prior fine-tuning to set up or prepare the way. Therefore, they are back to the same problem, which is who/what did the fine-tuning? It's a circle that won't end. When the improbability of evolution is too much to overcome, then often the evolutionist will resort to a redirect away from their failed hypothesis by posing a question, "Where did your God come from—who created Her?"

The Bible records that God is eternal. Doesn't this violate science that nothing can come from nothing? No. It is not that God came from nothing, it is that God is infinite, eternally existent. Infinity is an acceptable term in science, but it does require a measure of faith. The amount of faith required is inversely proportional to the amount of knowledge. The more knowledge, the less faith required. For example, to corroborate this claim and reduce the amount of faith required to accept this claim, the evidence to authenticate the Bible's claim of God being eternal is the Bible itself. Its divine accuracy is tested and confirmed by archeology and observed with its harmony with the sciences. The Bible has endured the toughest of scrutiny from the most ardent opponents, and still the Bible is without contradiction, without error, without violations in the sciences. Only divine inspiration could have 66 different authors spanning three thousand years writing in perfect harmony of doctrine, biology, modern medical science, cosmology, and so forth. The more than two thousand fulfilled prophesies authenticate that something beyond human ability wrote the Bible. And the apostles who committed their lives to follow Jesus illustrate they saw many miracles. Since Jesus required celibacy or monogamy, acknowledged that He was poor and His followers would not get rich following Him, and promised persecution, then this means that the apostles followed Jesus knowing they were not going to get women, fame, or money, but were promised persecution. And not to be superficial, but Isaiah 53:2 records (~700 BC) that Jesus was not good looking, and He had no physical form that would attract people. The only way to get them to follow Jesus was for Him to perform miracles and rise from the grave. If Jesus couldn't substantiate His claims with miracles, then the apostles would have stopped following Him. If Jesus fooled them with some fake miracles, then upon His death on the cross the apostles would have been scattered and given up because of the life of no women, no fame, no money, and promised persecution. And that is exactly what happened when Jesus died. The apostles scattered and gave up. Thus, only a resurrected Jesus, after being dead for three days, could have brought the apostles back to following Him in a life of no women, no fame, no money, and promised persecution. The only conclusion is that Jesus rose from the grave, and only God could have performed this miracle. Therefore, combining the evidence of the perfect accuracy of doctrine from cover to cover, the fulfilled prophesies, the harmony with the sciences, and the lives of the apostles and Jesus' resurrection, leads to knowing that when God claims He is eternal and infinite it is truth. The faith required to accept this is reduced as knowledge increases.

How does the Bible handle the beginning of information and life? The Bible records via the testimony of God, written by the hand of Moses, under the inspiration of God the Spirit, that God created life from Himself. Luke 3:38 records that Adam was begotten by God with "Adam, the son of God." God spoke life into existence with "God formed man of dust from the ground, and breathed into his nostrils the Breath of **Life**; and man became a **living** soul (Gen. 2:7)." Remember that *life* and *living* are represented by the same Hebrew word, *chay*. The Bible is repeatedly and explicitly saying "life came from life." This is why we were created in the image and likeness of God (Gen. 1:26), because God created us from Himself. This also explains that the information in the DNA code came from the very essence of God. Therefore, our DNA code came from parts of God's DNA code, and since God created all the other things from His essence, then they too have parts of His DNA code. Thus, there are similarities in the DNA codes between creatures.

Does the Bible violate the scientific method? The Bible states that the life of everything living came from God Himself, which is in harmony with the science that shows that life doesn't come from nonliving material, just as Louis Pasteur proved in the late 1800s. But where is the observable portion of God? Stephen Meyer is on the cutting edge of this topic. Besides *Signature in the Cell*, I also recommend that you read *Darwin's Doubt*.

Man has spent billions of dollars and countless hours searching the heavens for evidence of life to potentially explain how life began on earth. What are some of the indications in the heavens that mankind is looking for? Astronomers look for anything and everything, from complex machinery to straight lines to curved lines. Can a straight line or curved line indicate life? Yes. How? The existence of life often includes things in geometric shapes, such as parallel straight lines, circles, and triangles. If cosmologists find a straight line or circle, this indicates intelligence. Something or someone with intelligence put things together to form a geometric shape. This isn't restricted to space; the same application applies to almost every science. When archeologists stumble upon a circle of rocks, they know that intelligent life did it. When archeologists or geologists see geometric shapes after digging down into the ancient past, they conclude that intelligent life formed the geometric shapes. They never assume that stones in a circle can evolve into a circle given enough time.

Whenever we see a watch on the ground, we know that a watchmaker (intelligence) made the watch. It doesn't matter if we found the watch in the ocean or in the desert, we would come to the same conclusion. Everyone knows that a watch can't form by itself, even if it has an infinite amount of time. Whenever we see a cotton shirt, we know there was a designer, an intelligent being that created the clothes. No one on the face of the earth believes it's possible that a tornado moving over a field of cotton crops, even given 4.6 billion years, will generate a cotton shirt. Whenever we see anything and everything on earth that has a design indicating a designer and a function indicating intelligence, we know there was a designer. We know someone created it and that there was an intelligent being behind the creation. No one believes that a printing factory can explode and produce a new work of literature, even if we exploded a billion factories a billion times a day for a billion years. The odds are zero.

The observable evidence is slowly taking form. When we observe items that have a design, function, and a purpose, we know by repetition, pattern, and history that there had to have been a designer and a creator behind the item. And that item didn't evolve by random chance or because of the wind, erosion, an explosion, a primordial soup of complex chemicals, or any amount of time. This is our first clue toward an intelligent designer behind the design of humans. But we aren't done.

Review: There is observable evidence of intelligent design in the things we see in life. We can test it and confirm it. When we see what seems like intelligent design, indeed there is an intelligent designer that created it.

Now, let's delve into the observable and testable portion of the scientific method to determine whether there was a designer and creator of human beings.

In life, we see that humans will do certain things to protect what they feel is precious to them, especially when it comes to things they created. For example, when sculptors create something, they put their seal on the image/sculpture with a signature to protect their creation and let everyone know it's theirs. When authors write something, they put their seal on it by copyrighting it to protect their creation and let everyone know it's theirs. Inventors similarly patent their inventions, and parents give their children their name.

We are created in the image (shape) and likeness (character) of God (Gen. 1:26), and when we create something, we are performing the very same actions that God performed ~6,000 years ago when He created life after His own image by begetting Adam, His first physical son (Luke 3:38). How did God patent His invention? Did He sign His sculptures, copyright His information, and give His name to let everyone know that He is the designer and creator of mankind? Yes, through the command multiply *"after their kind."*

The human female ovum (egg) has a shell around it to seal and protect the chromosomes of the female from being fertilized by the wrong kind. This shell surrounding and protecting the ovum must be dissolved in order for fertilization to occur. For the shell to dissolve and create an opening, it needs a specific enzyme from an outside source (from the male) to come in contact with the ovum shell. The female ovum shell is receptor specific; in other words, it will only allow one type of enzyme from one kind of creature to dissolve the shell for fertilization. To make matters more exclusive, the means of contact for fertilization is only by means of a chromosome-carrying device that has its own source of locomotion and its own energy source. The ovum is deep within the body and protected by gravity and acidity. Spermatozoa have a flagellum tail for locomotion to swim against gravity; they utilize a high-pH serum to neutralizes the acidity of the initial reproductive tract of the female. In addition, multiple spermatozoa are needed to dissolve the ovum shell; in fact, hundreds of spermatozoa must reach the ovum to have a sufficient number of acrosome granules (enzymes) to dissolve the shell.

Well, that's where one of the seals of God comes into play. There is only one kind of creature on earth that has the proper enzyme to dissolve the human female ovum shell for fertilization. It's the human male. At the tip of the human male spermatozoa is an acrosome granule; its sole purpose is to dissolve the shell of the ovum for fertilization. The male spermatozoon is also receptor specific for only one kind of creature on earth and only one kind of ovum. No other tissue that the spermatozoa comes in contact with will cause the spermatozoon to release the acrosome granule prematurely. Only the human female ovum has the correct protein receptor sites to trigger a release mechanism on the spermatozoa to release the acrosome granules for shell

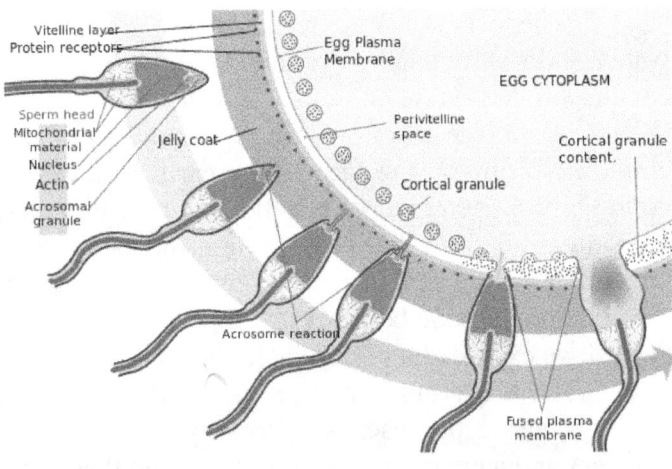

dissolving. Imagine the disaster if spermatozoa released the acrosome granules upon contacting any tissue. No fertilization would ever occur, and all kinds of creatures would become extinct in short order. *Image credit: Wikipedia.org.*

No other creature on earth besides the human male has this specific acrosome granule to dissolve the female human ovum shell for the fertilization of the chromosomes. And no other creature on earth besides the human female has the protein receptor-specific sites to trigger the male acrosomal enzyme release and accept the granules of the human male spermatozoa for fertilization. This is two-way specificity that establishes exclusivity. Don't be fooled by websites claiming that mankind has been cross-breeding different kinds for years; that is a switch of terms. Mankind has been cross-breeding different species of the same kind, but no one has cross-bred two different kinds of creatures.

So how is this a testable scenario that proves the hypothesis of God being the intelligent designer and creator of human beings? Well, it's a dubious test, but through the debauchery of mankind and the unfortunate sin of bestiality, mankind has on multiple occasions and over many millennia performed this "test" that determines if different kinds of creatures can crossbreed with the human kind. The answer is emphatically "No."

Since God created us in His image and likeness, He sealed us to protect us from ourselves in order to keep His creation in the form that He intended it. God gave Adam His name to claim him by claiming Adam as His son (Luke 3:38) and giving Adam the Breath of Life from the Holy Spirit. God made it so that each kind of creature could only reproduce with the same kind of creature. And that is why in the Gen. 1 creation account, one reads nine times the phrase "after their kind." God created each creature "after their kind" (Gen. 1:11, 12, 12, 21, 21, 24, 24, 25, and 25), from grass to plants, to trees, to fish, to birds, to beasts, to mankind; all are created after their kind by God to reproduce only with their own kind. All life on earth can only reproduce after their kind. This is an amazing seal and an amazing signature by God that reveals the exclusivity of His creation.

It should be noted that evolution by definition would not allow such exclusivity of kinds (kind is a larger category, while species are varieties within a kind) of creatures. Since evolution is based on random and unguided mutations and survival of the fittest, there wouldn't be exclusivity of kinds; quite the contrary, crossbreeding of kinds would be encouraged to determine the best kind of creature. Even the famed movie *Jurassic Park* has a line that says that one can't make evolution exclusive—evolution can't be boxed in. Evolution would not allow such exclusion of kinds. Evolution would find a way to cross the different kinds. Remember in the movie where the scientist of chaos Dr. Ian Malcom learns that the creatures in the park were all female dinosaurs so that the owners could control the population? Below are quotes from the movie.

John Hammond: I've been present for the birth of every little creature on this island.
Dr. Ian Malcolm: Surely not the ones that are bred in the wild?
Henry Wu: Actually they can't breed in the wild. Population control is one of our security precautions. **There's no unauthorized breeding in Jurassic Park.**
Dr. Ian Malcolm: **How do you know they can't breed?**
Henry Wu: Well, **because all the animals in Jurassic Park are female.** We've engineered them that way.
Dr. Ian Malcolm: But again, how do you know they're all female? Does somebody go out into the park and pull up the dinosaurs skirts?
Henry Wu: We control their chromosomes. It's really not that difficult. All vertebrate embryos are inherently female anyway; they just require an extra hormone given at the right developmental stage to make them male. We simply deny them that.
Dr. Ian Malcolm: John, **the kind of control you're attempting simply is, it's not possible. If there is one thing the history of evolution has taught us it's that life will not be contained. Life breaks**

free; it expands to new territories and crashes through barriers, painfully, maybe even dangerously, but, uh, well, there it is.
<u>John Hammond:</u> There it is.
<u>Henry Wu:</u> **You're implying that a group composed entirely of female animals will breed?**
<u>Dr. Ian Malcolm:</u> **No, I'm, I'm simply saying that life, uh finds a way.**

Dr. Ian Malcolm tells them that they can't confine evolution. Evolution will find a way. And guess what, evolution does find a way in the movie, and everything goes crazy shortly thereafter. Why? According to evolution, there is no exclusivity; nature will find a way to crossbreed kinds. Nature will find a way to reproduce new kinds of animals and a way to evolve new kinds. Transitional creatures that are half of one kind of animal and half of another should be normal to see around earth, not 100% of humans being humans, 100% of apes being apes, and 100% of each kind being the same kind. But there is no evidence in the past or present of any dog becoming any other animal or plant. Why? God sealed His creation from the beginning by coating female eggs and providing only the same kind of male creatures with the enzyme to dissolve the coating for fertilization. Also, from the beginning, God sealed the hormone receptor neurons of males to only be receptive to the same kind of female hormonal release of pheromones. Therefore, hormone production is to attract the same kind.

Arthur Conan Doyle's, Sherlock Holmes said it best, "Once you eliminate the impossible, whatever remains, however improbable, must be the truth." I am suggesting that the exclusivity of reproduction with only the same kind of creature, eliminates the evolutionary hypothesis of unguided randomness and only allows a creator hypothesis. Why? there is too much order and design governing the exclusivity of the reproductive process of all living things on Earth.

The old adage suggests that the glass is half full or half empty. The evidence of the Creator is out there for all to see, but how the evidence is interpreted is where the disconnect is. This is the heart of man interpreting the evidence that science uncovers; it's not science that is against God. It's this mindset in which so-called unbiased evolutionists interpret evidence and come up with theories that are anti-God. It's not that science is anti-God; quite the contrary, science is in harmony with the Bible.

By the way, it should be noted that each male dog has the correct enzyme for each female dog, from the wolf to a Maltese dog. And each male cat has the correct enzyme for each female cat, from the tiger to a tabby cat. Logistics make it unlikely that a male tabby cat could fertilize a female tiger, but it is possible. But no male cat has the correct enzyme to dissolve the shell to allow the fertilization of a female dog, even if the logistics were plausible. And it's like this for each creature. Each kind of male fish can only fertilize the same kind of female eggs, and each kind of male bird can only fertilize the same kind of female egg. Please note that a "kind" is a larger grouping of animals. For example, all breeds (varying species) of dogs descend from the wolf and are of one kind of animal.

Review: God sealed each creature to only be able to reproduce with its own kind. Medical science proves that there is exclusivity with reproduction and that each kind of creature can only reproduce after their own kind, which proves the Bible correct and proves evolution is wrong and in violation of medical science. This is testable evidence of the existence of God.

Evolutionists believe that mankind shed its fur because "6–8 million years ago apelike ancestors of modern humans had a semiaquatic lifestyle based on foraging for food in shallow waters" (Scientific American). Then, mankind had to kill other creatures so that they could put on heavy coats to survive the ice age. What an evolutionary bummer that we lost our fur heading into the ice age. The theory is that humans lost their fur so that they could swim better. And the proof of this is our remnant webbed hands and feet. This is a fanciful hypothesis that is faith based, yet it is acclaimed as fact.

This leads us into the discussion of vestigial organs. A vestigial organ is something that has no function and has lost its original purpose. Vestigial organs are offered as proof that we evolved from another creature. For example, the most famous alleged vestigial organ is our tailbone. Evolutionists cite the tailbone as proof that we once had a tail like our forefathers, the primates. Here are the problems with that: (1) Again, evolutionists want to show a loss of something to prove that we gain information. This is backwards thinking; why would we lose something that is functional? This is the fanciful dream world of evolution. They never see any human with a partial tail, but they assume we all had a tail millions of years ago. (2) The coccyx is not called a tailbone; it's called the coccyx because it resembles a cuckoo's beak, and it is not a remnant tail. It has a function and a purpose. The function is to allow the attachment of many muscles, ligaments, and tendons. The purpose is as a structural support for all the visceral organs, and more importantly, for the sphincter control of involuntary muscles, so we can defecate at the appropriate time. Without the coccyx, feces would just ooze out via peristaltic motion without control. Thus, if any evolutionist proclaims the coccyx to be a remnant tail, they do not know anatomy or anatomical functions.

The other famous vestigial organ is the appendix. The appendix allegedly proves that we evolved from another kind that once used the appendix. Since the appendix allegedly has no function and no purpose, it proves we evolved. Well, this is antiquated thinking, and frankly, it's a hypothesis based on ignorance. Medical science has finally caught up to the bizarre wild claims of evolution. Several decades ago, we discovered that the appendix stores bacteria that are necessary for the breakdown of food for fuel. When humans eat something bad, or when their bodies are overwhelmed with foreign bacteria that cause diarrhea, the healthy bacteria in the large intestines needed for the final breakdown of ingested food for nutrient absorption get flushed out of the intestines. Well, guess what? The appendix, which stores the natural, normal, and needed bacteria for the intestines, is there to save the day. The appendix releases stored bacteria into the intestinal system and reconstitutes the healthy, needed bacteria into the intestines. Without the appendix to save the day, humans would get diarrhea and die of dehydration. Therefore, the appendix is involved in immunity. Don't think for a second that the appendix has no function and no purpose. The appendix is not from an ancestral creature that we evolved from. Again, evolutionists want to use the logic of losing something, in this case the appendix, as evidence that we gain new function and new kinds via random, unguided mutations of the DNA code. It's illogical to show a loss of a function as proof that we gain new functions and new kinds. Whenever, an evolutionist proclaims the appendix proves humans evolved, know that the only thing their wild claims prove is that they don't know medical science.

As discussed previously, evolutionists use the term *pharyngeal gill slits* to describe the folds in the skin on a human embryo. They proclaim that these folds are vestigial in nature, indicating that man evolved from fish. Creationists and those familiar with anatomy know that each fold of the tissue of an embryo grows into a different aspect of the human nervous system; in no way do these folds in embryonic tissue represent that we once had gills from ancestral fish. This is really just a lack of knowledge on the part of those who suggest that the folds in the tissue are from prior gill slits.

I have even heard an evolutionists use the pineal gland as evidence of evolution because the gland has no function. Well, mankind has recently found out its function and that it's responsible for hormone production for such things as sleep inducement. The pineal gland produces melatonin, which is a serotonin-derived hormone that influences sleep. A full night's rest is important for decelerating the aging process. What was thought of as a useless vestigial organ, which supposedly proved evolution, actually aids in sleep regulation and youthfulness.

Well, what about wisdom teeth? Are they vestigial? No, they have a purpose. The purpose is to push out and force the other teeth together to close gaps when one tooth in front is knocked out or rotted out. Many people don't get their wisdom teeth extracted, and the teeth become useful in breaking down food for consumption. They are not vestigial at all. Wisdom teeth come in handy when humans live 900+ years.

And lastly, we come to the alleged vestigial thigh bone in whales. It is said that the bone is from when the whale used to have a leg and that it no longer has a function. On the contrary, the bone is used for mating and reproducing. It shouldn't even be called a femur; it is part of the pelvis. The label for the bone came from evolutionists trying to provide evidence to support their beliefs. The bottom line is that the bone is part of the pelvis and is useful for mating and reproducing offspring.

Medical science has caught up to the wild claims by evolutionists. Now you are more informed about these alleged vestigial organs that are really based on not knowing how the body works and forcing creative fanciful ideas to fit their beliefs. In fact, there is not one single vestigial organ on any creature alive or dead to show that any creature evolved into another kind or from another kind. There is no need to go through each fanciful tale of supposed vestigial organs; at some point, there is no logic to chasing an endless trail of wild claims. Let's just cut to the chase. There are no vestigial organs—each organ has a function and purpose that is accounted for. What is sad is that atheistic evolutionists are pushing hard on these wild claims, which are not sound in logic and based on not knowing anatomy, and yet theistic evolutionists buy into this illogical deduction and theories based without knowledge of anatomy that reject Biblical claims. Theistic evolutionists try to appease both sides in the war on truth, and God declares you are either 100% for Me or 100% against Me. You can't believe 50% of the Bible and 75% of evolution. They are exclusively on opposite sides of the war on truth that both sides are actively engaged in.

There is one animal that actually does have a partial psuedo-vestigial extremity, but it doesn't support evolution; instead, it authenticates the Bible. And that is the snake. The snake does have a rudimentary extremity that represents a time in the past when the snake once had full extremities and crawled on limbs instead of slithering on its belly. Gen. 3:14 explains that as a result of a particular snake allowing Satan to possess it and subsequently being an agent of the Devil to deceive Eve, the snake lost its ability to crawl on legs and was forced to slither on its belly. The snake still has small, leftover partial claws, from when it once had limbs. The snake still uses those claws for mating but no longer uses them for mobility.

There are three functions that existed at creation but have been removed because of sin. One is the legs of snakes that were once used for crawling. Yet, they still have a function, and that is for mating. The other two are not tangible things that we can observe: they are that potentially humans used to be able to see in the spiritual realm and animals used to be able to talk to humans. The last two are discussed toward the end of the book, but they deal with Balaam and his donkey in Numbers 22 and the snake in Gen. 3.

Review: Evolutionists' claim of vestigial organs that prove creatures evolve only prove evolutionists' lack of knowledge regarding the organs that they think have no function. Logically, the hypothesis of evolution does not get credit for showing something that allegedly has lost function to prove we gain new functions and new kinds. These are illogical, wild claims just to prove there is no God.

Someone reading may still be wondering what evolution is. To break down the theory of evolution into lay terminology versus Ernst Haeckel's "ontogeny recapitulates phylogeny," it is the compilation of tiny changes in organisms via mutated DNA over time, adapting to an ever-changing environment and stimuli, coupled with natural selection to assure that the best mutations survive. Those subtle and tiny changes accumulate over time and build on each other with each passing generation, and this is how life evolves from a single-cell organism to a human being after 1–2 billion years of evolution. It is illustrated by the following series of shapes as a crude example:

!ⵑĹŁḳḲλÝYỲVV̌∆◊⌂✺¤⌾OOʘꙨωꝏ

From the schematic above, you can see how an exclamation mark evolved into an infinity sign through subtle and accumulative changes over time. This sounds reasonable, and on the surface, it does make sense. Is this truly how it is? This is where the debate of evolution versus adaptation occurs. Adaptation (or speciation) is accepted by creationists as fact. And adaptation occurs in all life forms, from animals to botanical life. There are subtle changes due to adaptation to the environment, but the "kind" of creature remains the same. A creationist would say that the above illustration may start off as an exclamation mark, and it may become larger or bold, turn into italics, or change font, but in the end, it's still an exclamation mark, and can't evolve into an infinite symbol. For example:

!!*!*/*!*/*/!!*!*!*!*

The different types of exclamation marks represent species from adaptation, yet they are still the same "kind" because the DNA code for "kind" remains the same and the genome adapts with a different emphasis on the information dependent on combining parental DNA and environmental conditions. When a mutation occurs to the question mark and it loses its function, then it becomes an indistinguishable blob, and there is a loss of function or a loss of information, not an enhancement of information. Evolutionists always use adaptation to show that evolution is real, but adaptation and evolution are completely different. Adaptation is based on information already in the DNA code, and the DNA code limits how far a creature can adapt before death or sterility occurs. Adaptation already has defined limits set by the DNA code, and it's impossible to evolve to different kinds without new DNA coding that wasn't already there. It is analogous to a computer software program having changes done by random keystrokes by blindfolded monkeys who accidentally improve the software program, not just once for one benefit, but trillions of trillions of times. Evolution requires the DNA to get new information for the changes to occur for new function or new kind. The two (adaptation and evolution) are very different; one doesn't authenticate the other.

At the end of the day, at the end of the millennium, the creature is always the same kind, even though it has adapted to its environment. Since no one observes that adaptation is not limited and no one can observe the alleged evolutionary changes millions of years ago and since it's so slow today that no one can observe a change of kind today, that's where science ends and faith takes over.

A problem for evolutionists is the sudden explosion of complex and diverse creatures during a brief Cambrian period. Allegedly, evolution is very slow, however, during this time period there were new complex creatures that seemingly evolved in too short a time period given the slow hypothesis of evolution. A very clever evolutionist might explain this rapid evolution by demonstrating how a bipedal creature quickly grew a brain that was three times larger than a human brain two million years ago. Can biologists take a mammal, such as a mouse, and experiment with varying stimuli, such as protein inhibitors, beta genes, and so on, to explain the alleged accelerated brain growth of transitional Australopithecus to a more advanced Homo erectus? Is there any way to have a mouse's brain grow in size in a single step? It turns out that the beta-catenin gene, which coats a protein, helps cells regenerate and causes cells to get bigger. By regulating the beta-catenin gene in the central nervous system, biologists have inactivated the amino terminus, which is a component to keep the gene active. Then, they developed an enhancer and put it in the mouse and raised the mouse under those conditions. The result: the mouse brain was twice the mass, with increased sulci. Evolutionists say the mouse adapted and evolved a larger brain, just like earlier primates evolved to have larger craniums and to eventually become human. However, there was no evidence that this benefited the mouse, and the adverse effects from this experiment were not included in the data. This is just a mouse artificially adapting as though a DNA code was instructing the regulation of the beta-catenin gene, not a new function and not enhanced brain power. No information was increased in the DNA code, and no offspring were produced that had an increased brain size. This is not evidence for evolution; this is evidence of adaptation by intelligent design of the scientist. Now, if the mouse became even slightly smarter, or if the mouse's DNA added new beneficial information, then evolutionists would have something, but that is not what happened. Therefore, the test merely demonstrates intelligence design to artificially illustrate the ability of the body to adapt and not the ability of the body to evolve a new function.

Review: Adaptation does not lead to or authenticate evolution; they are exclusively different. Adaptation utilizes information in the DNA code to modify features, whereas evolution relies on random, unguided changes to the DNA code that produce new functions and eventually new kinds. Evolution is impossible, and evolutionists erroneously use adaptation to save them.

Take Darwin's Galapagos finches; they had DNA code allowing them to be finches with beaks. Through adaptation, their beaks slightly elongated by a couple of millimeters, but they were still finches. The DNA of a kind remained the same, but the genome got modified. The finches that had DNA from birth for elongated beaks that better suited that environment, thrived and dominated the population. This is exactly how every species of human, though sharing the same DNA from Adam and Eve, but because of environmental conditions, sin, and so forth, there are humans with different color of hair, shape of hair, color of eyes, color of skin, tall, short, and so forth.

And most importantly, the information to adapt to the external stimulus was already in the DNA code; it wasn't new information being added into the DNA to allow beak modification. In other words, take Adam and Eve, they both had the DNA information for all the differences we see today in humans. Yet, every human is the same kind, made in the image of God, sharing DNA with God, and God is their physical Father. That's a big difference from saying that random, unguided mutations altered the DNA, and this caused the beaks to alter shape—and those whose beaks were altered poorly died off, and those whose beaks were altered appropriately for the terrain survived.

Suppose that Noah took a set of male and female wolf pups on the ark. After millennia of breeding, migrating to different regions on the globe with different environmental conditions, and adaptations governed by existing DNA, we now have thousands of different species of dogs, from the wolf to white fluffy Maltese, to beagles, and so on. But they are still dogs. Why? There is a protective coating surrounding the egg of a female dog, and only one kind of creature on earth has the necessary enzyme at the tip of the male sperm to dissolve this coating for fertilization to occur. Guess what that creature is? A DOG! Again, why is this? Because God has sealed His creation from the beginning for plants and animals to produce after their kind. This exclusivity of breeding is for each kind of creature.

Also, in the hormonal makeup of each kind of animal, there are specific pheromones that attract only the same kind of animal. When a female dog goes in heat and she is ready for a mate, she releases pheromones to attract a mate. Why is it that other kinds of creatures don't come to fertilize her? Why don't humans get excited and court the female dog? Why doesn't a male cat get excited and court the female dog? Because God has sealed His creation so that all plants and animals produce after their kind. In fact, the pheromones the female dog releases to attract a mate has no biological affect on other kinds of animals, except the male dog. This deals with DNA. No other animal has the appropriate protein receptors that are specific to receive the female dog's pheromones and translate the stimulus into a mutual hormonal mating response, except a dog. This is similar for every animal, fish, bird, and human. That is why no crossbreeding of kinds is possible and explains why 100% of dogs are 100% dogs, and 100% of humans are 100% humans, and so forth.

Evolution never knows where it's going; it's a free-running, random, unguided program, with survival of the fittest kind of guiding its steps along the way. By definition, there should still be chimpanzees making evolutionary steps and jumps, and survival of the fittest should either allow new transitions to succeed or eliminate them. We should be able to observe a human devolving back to a chimp because of unguided, random mutations, and we should be able to see chimps evolving. But we don't. Evolutionists conveniently say that the process is too slow to see. The changes occurred hundreds of millions of years ago, yet evolutionists don't know that. Each day, each year there should still be those changes that evolutionists proclaimed happened 300 million years ago. It's a nice way to hide their lack of evidence.

Evolutionists might say, "Hey a horse and a donkey produce a mule, and a lion and tiger produce a liger." Yes, but they are the same kind. A horse, donkey, and mule are all the same kind, though different species. And the mule is sterile. Likewise, the lion, tiger, and liger are all the same kind, though different species. And the liger is sterile. Thus, there are limitations to how far a creature can adapt before the DNA ends the adaptation with such things as sterility. This is observable evidence corroborating the Bible's version of creatures can only produce after their kind, and debunks evolution.

When did God seal the grass to remain as grass and the plants to remain as plants and the trees to remain as trees? It was on the third day of creation with the phrase "after their kind" in Gen. 1:11–13. God repeats this phrase three times to drum home the point for botanical life. Why? Because plants needed to be told? No. Because God knows the mind of man. God knew man would think that mankind evolved from other created things. Darwin thought he was the first to come up with the idea that we all evolved from rock minerals, trees, and the like. He was wrong. Consider Jeremiah 2:27:

> Who say to a tree, "You are my father," And to a stone, "You gave me birth." For they have turned their back to Me, And not their face; But in the time of their trouble they will say, "Arise and save us." But where are your gods Which you made for yourself? Let them arise, if they can save you in the time of your trouble.

You see, this is an age-old tactic to remove God as creator. To say we evolved from a rock is to say that one rejects the Words of God. And Darwin is not the originator of this movement. Nor is any other human for that matter. Now wait a minute; no one thinks they literally came from rocks, do they? Well, let's trace the evolutionary tree backward. Where did humans come from? Primates (apes, monkeys, and so forth)—and where did they come from? You keep going backward until you get to the beginning, where it rained on rocks to create a primordial soup of chemicals. Voila, rocks are the starting point. If evolution is false and deceitful, then it may be from deceitful spirits and thereby a doctrine of demons (I Timothy 4:1). Evolutionists would be wise to ponder the implications if this be true—that the concept of evolution is purely an attempt by Satan to keep humans away from the Word of God, thus trapping them in the slavery of sin and eternal death.

Review: Evolution violates medical science with the belief of crossbreeding of different kinds, which is proven to be impossible by the exclusivity of the fertilization process. Evolution violates the scientific method by believing in something that is not observable and not testable, with notions such as the spontaneous spawning of first life, the spontaneous forming of DNA coding, mutations enhancing the genetic code, and on and on. Evolution violates logic with a bait-and-switch scheme by noting what is observable, such as adaptation, and using that to suggest that what is not observable is true. Evolution violates logic by vestigial organs as an apparent lost function to demonstrate that we gained new function, and each reference to an alleged vestigial organ only points out the lack of knowledge of anatomy.

Group Discussion:

1. What had the most impact on you in this chapter?

2. Have you noticed that science is an ally to the Bible, and that it is only evolutionary scientists with their interpretive spins on the evidence that is contrary to the Bible?

Chapter 31
Evolution Versus the Bible

There are many Christians who believe in evolution, but they suggest that God performed or initiated the process. It's God's creative act that allowed the first single-cell life form to exist and to eventually evolve into all the complexities of life via mutations and natural processes. Can the two be in harmony? Can a believer in God believe that God orchestrated the evolutionary process over 14 billion years? Or does a theistic evolutionist's beliefs lie in complete conflict with the Bible? What does the Bible say about evolution, or is the Bible mute about the subject?

The Bible puts the death nail in the concept of evolution in Romans 5:12, "Therefore, just as through one man sin entered into the world, and death through sin, and so death spread to all men, because all sinned . . . death reigned from Adam." Since there was no death in the world before Adam sinned, then there cannot be one fossil before Adam sinned. And since the notion of evolution is that through a series of failed attempts, death, and fossils, then man evolved. Then God completely obliterates the notion of evolution by saying that there was no death in the world before Adam sinned. Furthermore, God says in Jeremiah 2:26–28:

> As the thief is shamed when he is discovered, so the house of Israel is shamed; They, their kings, their princes and their priests and their prophets, who say to a tree, "You are my father," and to a stone, "You gave me birth." For they have turned their back to Me, and not their face; But in the time of their trouble they will say, "Arise and save us." But where are your gods which you made for yourself? Let them arise, if they can save you. In the time of your trouble.

These Jews were theistic evolutionists, and God declared they were committing idolatry and were shamed. Even Jesus quotes Genesis, indicating that He believed in a literal interpretation of creation, that God made mankind male and female from the beginning. Jesus is so stern about this that He questions them with "Have you not read?" Today, those who want to believe that man evolved by evolutionary means are faced with Jesus' own words, (Matthew 19:4-6) "Have you not read that He who created from the beginning made them male and female, and said, 'For this reason a man shall leave his father and mother and be joined to his wife, and the two shall become one flesh'?" For those who think they can believe in God and evolution, "Have you not read?" Jesus 100% ruled out the possibility that God created a single-cell prokaryote that evolved into other life forms. Jesus 100% closed the door on man evolving from evolutionary means.

Yet God is merciful, and only God knows when someone has a true repentant heart. He knows that people have been taught—from all of academia to every television show—is that evolution is fact. God has special verses for those who are without knowledge of the Bible or reject the Bible and the notion that every mountain was covered under the heavens for a global flood (Gen. 7:19–23) and for those who are without knowledge of or reject a seven-day creation and accept a re-creation or evolution (Exodus 20:11), which make them guilty of annulling or setting aside or modifying or neglecting some of Scripture. A believer of Jesus cannot lose their salvation (Rom. 8:38–39) and will enter the Kingdom of Heaven, but those who set aside some Scriptures to accept evolution will be called least in the Kingdom of Heaven (Matthew 5:19). Those who hold the Scriptures in low esteem will be held in low esteem by Jesus. A harsher warning for those that lead (teach) people astray with false doctrine, such as with evolution (Matthew 18:6).

Review: A believer in God won't call His Word a lie or inaccurate because Jesus is the Word. It is possible for a believer to make errors, such as believing in evolution, but once this information has been brought to a true believer's attention, they will mourn for calling God a liar and reject

evolution because God rejected evolution in Jeremiah 2:26–28 and Matthew 19:4–7.

Some people believe in the singularity of the Big Bang theory and God, but the two concepts are mutually exclusive because of the premise that before the Big Bang, 14.6 billion years ago there was a singularity. The Big Bang theory suggests that all that exists in the universe was compressed into a tiny dot as energy. Yet before God expanded the universe and the atmosphere on the second day (i.e., the Big Bang), He created all the matter on the first day. This means that all the matter created on the first day of creation would be outside of the small dot-sized universe. Also, on the first day of creation, there were waters, and all matter was formless, which excludes the dot size form of the singularity. Either the Bible is wrong by saying that the matter in the universe was formless, or Big Bang theorists are wrong by saying that energy had the form of a dot. Also, the Bible records "waters" before the Big Bang, indicating that the volume of the universe was small before the violent expansion, but not too small to prevent water and dirt from existing. Just prior to the Big Bang, at the singularity of the dot, water couldn't exist according to the Big Bang theorists because space, heat, and energy consumption would have prevented the existence of water, and all matter existed as energy, not water. But the Bible is clear that waters existed at the beginning before the Big Bang.

A big problem for humans is that they are being staunchly told that the earth is billions of years old by seemingly the smartest people on the planet who use sophisticated science to allegedly prove it, with such things as radioactive decay dating methods. Most people think that scientists are telling them the truth, which makes God's Word non-truth. Why do evolutionists take the leap of faith and believe that the earth is billions of years old? The evolutionist doesn't accept the notion that they believe in anything. They say it's fact, and radioactive dating is just one part of the proof. But they don't know or won't believe that the foundation for radioactive dating is based on a "Constant Rate Decay," which must always be constant. We have shown that the CRD is not constant. There are many problems with evolution, and the gaps between Darwinian evolution and what we know to be truth are getting larger, so why are many evolutionary scientists and believers still adhering to their beliefs in evolution?

It is fair to argue that errors made in the beginning of a theory that have other ideas built upon them will cause greater and greater error margins with each chain of hypotheses built upon the erroneous foundation. And that is what creationists say is wrong with the hypothesis of evolution; it starts wrong with the origin of life and the origin of information and increasingly gets further from the truth with each progressive thought built upon the foundational errors. And this sums up the entire evolutionary model: a set of errors built upon initial error.

From a medical perspective, the hypothesis of evolution is like a sick patient with increasing symptoms and different ailments that are becoming more chronic, with new and more acute symptoms flaring up with each increase in mankind's knowledge of life and understanding of the universe. What is the root cause of why people believe in evolution? There are two answers: the voluntary realm and the involuntary realm.

The voluntary realm.

For the evolutionist: This is a willful attempt to justify sin. Take the racially motivated Darwin for example; he believed he was of a superior race, and this is sin because the Bible explains we are all created in the image of God, and it is a race to be the humblest, not a race to be the most superior. Darwin already accepted the sin of racism and interpreted empirical data based on this foundation. Thus, Darwin attempted to justify his superiority and that his species had evolved further than other human species of color—through evolution. Therefore, evolution was to justify the sin of racism. However, not all evolutionists are racists, and for this reason the voluntary realm needs a broader explanation, such as by drawing conclusions from the empirical data that fit their own beliefs as the standard or foundation on which to formulate such conclusions by conscious choice. For example, when one believes humans eventually evolved from a single-cell organism and fish was one of the evolutionary steps, then when they sees folds of skin on a human embryo, then the folds look like gill slits from when our ancestors were fish. This is a voluntary choice to interpret data that fits a preconceived belief. And that belief is that there is no God, and thus morality is relative. And for this reason, there is a connection with a voluntary choice to interpret the observable evidence so that it aligns with allowing one to live without the absolute morality of the Bible. The root is that there is no God, and the data is interpreted to support that conclusion. Thus, the result is denying God the honor due Him and giving man the highest rank by evolutionary means. And for this reason, evolution is linked to self-worship, human glory, and sin. Scripture addresses this with Romans 1:21–23:

> For even though they knew God, they did not honor Him as God or give thanks, but they became futile in their speculations [about creation], and their foolish heart was darkened. Professing to be wise, they became fools, and exchanged the glory of the incorruptible God for an image in the form of corruptible man and of birds and four-footed animals and crawling creatures.

For the Creationist: This is drawing conclusions from the empirical data using God's wisdom as the standard or foundation on which to formulate such conclusions by conscious choice. Proverbs 3:5–6: "Trust in the LORD with all your heart, and do not lean on your own understanding. In all your ways acknowledge Him, and He will make your paths straight. Do not be wise in your own eyes." Creationists accept the Word of God, which says that all life came from God as He wrote it through the inspirational writings of Moses. Therefore, this is a voluntary choice to interpret data that fits a preconceived belief. And that belief is that there is a God, and thus morality is absolute. And for this reason, there is a connection with a voluntary choice to interpret the observable evidence so that it aligns with allowing one to live with the absolute morality of the Bible. The root is that there is a conscious choice to interpret data to align with the Bible and give God the honor due Him.

Subsequently, both sides attain general conclusions from a conscious choice of whether there is or there is not a God. This choice results in a subjective bias that favors their already preconceived notions. This is a voluntary choice to form conclusions from viewing the physical evidence that fits what one wants to believe is the truth. It seems reasonable to draw conclusions based on what is seen and observed to determine what occurred at some distant moment in the past. But no one was there when these major earth-changing events occurred, except one, God (according to the Bible). Therefore, some form of faith-based decision making is required for both the evolutionist and creationist. Evolutionists see many indicators that may be interpreted as the earth being billions of years old; thus, they align with the conclusion that the Bible must have errors regarding the creation account. Therefore, they determine that the Bible was not supernaturally written, but written by humans alone. The problem is that evolutionists already come to the table with a notion that the Bible can't be true and base their deductions on that foundation and find data/numbers to support their view. Creationists abide by the very same principle when viewing the same information as evolutionists but interpret the observable through the eyes of the Bible.

Both sides make the same conscious decision to validate their prior convictions. The evolutionist consciously wants to disprove the literal creation account in Genesis, while the creationist wants to validate the Bible. For this discussion, we'll use as example the validity of the CRD in the formula for radioactive dating. Why do the evolutionists have a steadfast faith that the CRD has always been constant, when proving that the CRD is not constant is easy to do? They voluntarily want conclusions that support an old earth and come up with techniques (arithmetic fiction) that give credibility to an old earth, which supports their already-established faith that the Bible is unreliable. There is a deep methodology in evolutionary naturalism and reasoning that suggests that if one is going to be a scientist or call oneself a scientist, he or she must limit the reasoning for explaining processes to "chance," random mutations, survival of the fittest, and time. This is a deep-rooted conscious decision (and this creates subjective discernment that trumps objectivity) to interpret empirical data so that the Bible has some errors. And thus, they rely on their own wisdom to discern the evidence rather than the wisdom of God to determine the proper way to interpret the evidence. Evolutionary hypotheses require lots of time, so all interpretations are required to support this preconceived belief.

This voluntary aspect of discerning an old earth seems to be based on objective findings on the surface. But the quasi-objective evaluation and objective conclusions are only lip service to veil the subjective bias of the heart and mind. Evolutionists will tell you that they alone possess the objective mind to interpret evidence and that their data proves a very old universe.

What the evolutionists won't admit or can't admit is that there is plenty of evidence that the evolutionary hypothesis has failed the observable and testable tenets of the scientific method and that there is ample, compelling evidence of a young earth. They won't acknowledge that there are plausible young earth explanations for all the old earth theories of evolution.

Believers who accept mankind's interpretation of evidence (i.e., evolution) versus accepting the Bible as the standard of truth for which all science is measured against are guilty of one or more of the following: not fully trusting God, putting some faith in man over God, acting on a selfish desire to avoid looking foolish, or not diligently reading the Bible. They also may be guilty of not sharing the Bible with others, even though it commands us to share the good news about the Bible. Well, if they believe that the Bible is not divinely inspired, then there is less compulsion to read or risk embarrassment with sharing a man-made book. Also, they are free to do as they wish if they believe that the Bible is not written by God. In that case, God would not be monitoring mankind to check on the required obedience.

Review: There is a voluntary choice to interpret data to align with preconceived beliefs. For creationists, this is a conscious choice to honor God as creator, that His Word is true, and to follow His standard of absolute morality. For evolutionists, such as Darwin, his preconceived belief was racism. Through the sin of his superiority, he interpreted data to justify his sin. For you it may be different, but for all evolutionists, the preconceived belief is that the Bible contains errors, and/or that there is no God. And this allows for relative morality, and self-glory as the pinnacle of evolutionary achievement. This denies God what is rightfully due Him—honor as creator. Why believers accept evolution: they have a lack of knowledge about the Bible, of knowledge about evolution, and of faith in God. They fear mankind and prefer self-preservation over honoring God at all cost.

The involuntary realm

For the evolutionist: This is drawing conclusions from the empirical data using their own desire as the standard or foundation on which to formulate such conclusions by unconscious drive. There is no choice here; this is from within and ingrained in the inner person, from the heart. This is addressed in Colossians 1:21: "And although you were formerly alienated and hostile in mind, engaged in evil deeds." It's the age-old discussion of Martin Luther and Jonathon Edwards about the free will to choose or not. The answer is that we have free will to choose, but prior to salvation, all of our choices will be to glorify ourselves, satisfy our flesh, or test God. We have a free will to choose but can only chose evil because we are alienated from God, we have a hostile mind toward God, and our actions are evil self-glory. To compound the problem, evolutionists do not accept the Genesis creation account because it is foolishness to them, and they cannot understand it because their spiritual nature is contrary to the spiritual nature of God. I Corinthians 2:14: "But a natural man (not saved) does not accept the things of the Spirit of God, for they are foolishness to him; and he cannot understand them, because they are

<u>spiritually appraised."</u>

For the creationist: This is drawing conclusions from the empirical data using God as the standard or foundation on which to formulate such conclusions by unconscious drive. There is no choice here; this is innately ingrained in the inner person. This is addressed in the Scriptures with II Corinthians 5:14–18:

> For the Love of Christ controls us . . . we live for him who died and rose again . . . Therefore from now on we recognize no one according to the flesh . . . Therefore if anyone is in Christ, he is a new creature; the old things passed away; behold, new things have come. Now all these things are from God.

I Corinthians 2:7–16 says, "We speak God's wisdom . . . For to us God revealed them through the Spirit; For the Spirit searches all things, even the depths of God . . . We have the mind of Christ."

This is a spiritual motivation, an inner drive that directs and battles the conscious mind for power/control. It is a preconceived paradigm that forces us to configure our numbers and time scale to fit and affirm what we believe is truth. On the other hand, atheistic evolutionists suggest that there is no God and therefore no judgment, thus allowing them to potentially fulfill any carnal desire they want. They have an involuntary urge to lean on their own understanding and in all their ways to acknowledge that there is no God. The creationist does the same thing, but inversely, and suggests that there is a God and therefore a judgment, thus allowing freedom from the sins of our carnal desire and instilling a desire to live according to the Bible's moral code. Creationists have an involuntary urge to lean not on their own understanding and in all their ways to acknowledge God.

This involuntary response has a spiritual motivation. We are either enemies of God or an ally. Jesus said that you are either for me or against me. There is no middle ground (Matthew 12:30). Therefore, one either acknowledges God in all their ways, or they don't. If you claim to be a believer, yet reject what God wrote as truth, this is a good time to do a self-evaluation.

Review: There is an involuntary drive that affects the decision-making process. This is the condition of the spirit that affects the desire, thoughts, and motivation of the soul, and this affects the actions of the body. This is why we do not wrestle with flesh and blood (soul), but with principalities of the spiritual realm (Ephesians 6:10–17). The spirit of a saved person will have the mind of Christ and have the Holy Spirit to guide them in interpreting data and understanding the Bible. The spirit of the unsaved person cannot understand the Bible because it is spiritually discerned only through the Holy Spirit, and thus they are left spiritually blind and cannot see the correct interpretation of the physical realm.

Group Discussion:

1. Do you view the observable evidence through the lens of man's philosophies, or through the lens of Scripture? And what are the implications of which lens one views the observable for interpretations?

2. Read Deuteronomy 4:32 and discuss how this moves you away from evolution and toward Scripture.

If you have comments or questions to the author, email them to: Lawrence@creationministry.org.
If you wish to donate to this ministry, which is tax deductible, go to: www.creationministry.org.

www.ingramcontent.com/pod-product-compliance
Lightning Source LLC
Chambersburg PA
CBHW081013040426
42444CB00014B/3194